Changes in
Plain Bearing Technology

Other SAE books of interest

Engine Failure Analysis: Internal Combustion Engine Failures and their Causes
By Stefan Zima and Ernst Greuter
(Product Code: R-320)

Modern Engine Technology from A to Z
By Richard Van Basshuysen and Fred Schaefer
(Product Code: R-373)

Design Practices: Passenger Car Automatic Transmissions
(Product Code: AE-29)

**For more information
or to order a book, contact:**

SAE International
400 Commonwealth Drive
Warrendale, PA 15096-0001 USA

Phone: 877-606-7323 (U.S. and Canada only) or
724-776-4970 (outside U.S. and Canada)
Fax: 724-776-0790;

Email: CustomerService@sae.org;
Website: books.sae.org

Changes in
Plain Bearing
Technology

By Rolf Koring

Translation by Phil Dando

Warrendale, Pennsylvania
USA

400 Commonwealth Drive
Warrendale, PA 15096-0001 USA

E-mail: CustomerService@sae.org
Phone: 877-606-7323 (inside USA and Canada)
            724-776-4970 (outside USA)
Fax: 724-776-0790

ISBN 978-0-7680-7724-7
SAE Order No. R-420

Library of Congress Cataloging-in-Publication Data

Koring, Rolf.
   [Gleitlagertechnik im Wandel. English]
   Changes in plain bearing technology / by Rolf Koring ; translation by Phil Dando.
      pages cm
   "SAE Order Number R-420."
   Includes bibliographical references and index.
   ISBN 978-0-7680-7724-7 (alk. paper)
   1. Plain bearings (Machinery)--Technological innovations.
   2. Fluid-film bearings--Materials. 3. Babbitt metal. I. Title.
   TJ1063.K67 2013
   621.8'22--dc23   2012033050

Information contained in this work has been obtained by SAE International from sources believed to be reliable. However, neither SAE International nor its authors guarantee the accuracy or completeness of any information published herein and neither SAE International nor its authors shall be responsible for any errors, omissions, or damages arising out of use of this information. This work is published with the understanding that SAE International and its authors are supplying information, but are not attempting to render engineering or other professional services. If such services are required, the assistance of an appropriate professional should be sought.

To purchase bulk quantities, please contact:
SAE Customer Service
E-mail: CustomerService@sae.org
Phone: 877-606-7323 *(inside USA and Canada)*
            724-776-4970 *(outside USA)*
Fax: 724-776-0790

Visit the SAE Bookstore at
books.sae.org

The cover photo shows a tilting pad journal bearing from John Crane Bearing Technology GmbH, Göttingen, used with permission.
Translated from the German language edition: *Gleitlagertechnik im Wandel* by Rolf Koring
Copyright © expert verlag, Renningen, Germany, 2012.

Table of Contents

Preface

I was determined that before the end of my professional life I would document my experience in the field of plain bearings. While this is limited to metallic hydrodynamic bearings, it is no depreciation of other bearings such as low-maintenance, sintered, and synthetic bearings. It just happens that my experience is mainly limited to the metallic hydrodynamic bearings, and I want to write on a subject where I have confidence in my experience. My involvement with bearing technology started relatively late. It accounts for the second half of my professional life. As a result, maybe I am at a disadvantage compared with colleagues who did this job their whole lives. But I am convinced that my previous work and associated experience are very helpful in my present position. A person changing careers can bring a fresh and impartial approach to problems. Established "facts" will be questioned, and the routine of structured working always proves to be helpful.

After my academic studies and graduation as an engineer, for the first two years I made static calculations for steel bridges. Then for fifteen years I performed static and dynamic calculation of power plant steel structures. On the construction sites many defects occurred, and so increasingly the refurbishment of damages became my speciality. In those days I used to jokingly say of my work, "everything that fractures will be refurbished by the fracture mechanic."

All of this does not directly relate to plain bearing technology, but I noticed very early in connection with those activities that, in general in the industry, more money and effort will be spent on selectively repairing damages than on eliminating the root causes. I never was comfortable with these conditions, but at that time I was not in the position to change the situation. Already my endeavors to indicate the source of problems showed that I would make few friends that way.

So I remained hopeful of sometime getting a position where I had more influence on the general context. The realization of this hope by joining Th. Goldschmidt bearing technology was really good fortune. From the beginning I was multilaterally involved in metallurgy, production, application, technical support, and sales. That was for many years unchanged but, since the transfer of the bearing technology unit to the ECKA Granules group in 1998 and the acquisition of HOYT bearing metallurgy in 2004, the scope of my work has been extended enormously. As the general manager for all these activities, I gained increasing influence on R&D.

In the field of plain bearing technology, experts of many different disciplines work successfully. However, it can be observed that there is too little attention to the general problem and its solution, echoing my observations in my previous job. The phenomenon always happens when the general solution needs the cooperation of different disciplines. Interdisciplinary coordination becomes necessary but, in practice, very often this is not recognized or is ignored.

Nobody can claim to be an expert in all disciplines. That applies to me as well. With my multifunctional work, I came into a position to contact many experts whose work is bearing technology. From them I learned much. They raised my awareness for the total interactions.

I owe my basic knowledge of bearing metallurgy to Mr. Hilgers, my predecessor at Th. Goldschmidt AG. He had contact with Prof. Macherauch in Karlsruhe University and his assistant Dr. Eifler, later a Professor for metallurgy at Essen University and Kaiserslautern University. He taught me many aspects of metallurgy in detail.

On my R&D project for developing improved and environmentally friendly white metal bearing alloys, Mr. Bohsmann of the Max Planck Institute Stuttgart was a big help.

Many ideas and contributions also came from Professor Frommeyer, MPI in Düsseldorf. Through the years we took opportunities for spontaneous short meetings. Each time I came away with new input for further R&D activities.

I must also mention Professor Spiegel. He was involved for all his working life with bearing technology and tribology; his work is precise, authentic, and efficient. In particular, he has the ability to reduce problems to their basics and present them in a simple and understandable format. For his calculations and technical support of my newly developed fatigue test bench, I am deeply grateful. For intensive support with FEM calculations, I thank Mr. Schmitz from Renk AG and Mr. Persson from MAN Diesel. Lech Moczulski from MAN Diesel impressed me with his enthusiasm for laser welding and his many ideas, together with his colleagues Vagn Sture Hansen and Jesper Vejlø Carstensen. Mr. Jasnau from SLV Rostock and his team also provided good input to the HERCULES B project. With Mr. Luchner from BMW, I had some interesting discussions about laser lining.

In 2002, I had the chance to found the VDMA Group for plain bearing working. Most of the German bearing producers became members, and we all stay active with exchange of knowledge and experience. This wealth of input is reinforced by my worldwide contacts through my job in ECKA Granules.

It is impossible to name them all, but representatives of the association of bearing producers I want to mention here are Mr. Kowollik from GLS, the Edelmann brothers from the Edelmann company, and Dr. Hermes from John Crane Bearing technology. They are all very critical but constructive discussion partners.

Besides the bearing producers as direct customers, I also support many end users of bearings. In most cases, the contact first arises in connection with operating trouble in the shape of plain bearing damage and is followed by the necessary damage analysis made by me. That provides an extensive insight into operating procedures, especially of power plants. In this respect I want to mention Mr. Jungbauer from Ennskraftwerke AG and Mr. Maldet from Tiroler Wasserkraftwerke AG, as notably genial and inspiring discussion partners.

Finally, I come to Phil Dando. He became my colleague with the acquisition of British company Hoyt Darchem. I am very grateful for his active support in translating this book, even after his retirement.

With my collaboration on different working groups of ISO TC 123, I came in contact with further experts of bearing technology. Exhaustive discussions resulted eventually in practical standards. I thank the Japanese Delegation for good and constructive cooperation, especially Professor Someya from The University of Tokyo.

I must not omit my thanks to the owners of ECKA Granulate Group, Mr. and Mrs. Rohrseitz. They always supported my substantial R&D projects with commendable patience.

Worldwide contact with bearing producers and bearing users enabled me to continually expand my knowledge. The multifunctional job gave me the ability to always keep in sight a broad view of the entire picture. I learned that for a successful solution, it is important to recognize at an early stage all deviations of market requirements and to adapt the R&D activities to that. It is pointless to develop ingenious things for their own sake. The market needs must be the driving forces.

My motto is not to re-invent the wheel again and again. New expertise should be acquired to become a basis for further development.

So I want to bring my experience into this book. Following the title, it will cover the total range of topics from design to materials, production methods, quality, and operation, and ultimately the failures and damages of plain bearings. At many steps along the way, there will be hints on problems, insufficiencies, and also on R&D requirements. It is a historical review with all interconnections, but also an outlook into the future that has already begun for bearing technology.

Rolf Koring

Chapter 1

Introduction

During the past 100 years, all sectors of industry have expanded their knowledge and introduced many improvements. Plain bearing technology is no exception.

With each intended change, whether to design, materials, or operational conditions, all basic parameters have to be checked for validity and accuracy. That sounds logical and straightforward.

In practice, there is the problem that possibly not all conditions that should be considered have been identified. This can happen, especially when the conditions relate to other disciplines. It may thus be impossible to define some conditions and to observe them. Most people have experienced the sudden awareness of influential factors which then seem to be obvious. One is left wondering why they have not been detected earlier. This book will give suggestions on many correlations.

There are many varieties of plain bearing designs. What are the essential characteristics?

Also, there are many possibilities for selecting the backing material. That leads to the question of suitability and whether the right material will always be chosen.

Many lining materials for plain bearings have been developed. This book concentrates on hydrodynamic bearings lined with white metals, also called Babbitts. More than 200 variations of these alloys based on lead or tin exist worldwide. Does the industry need such diversity? Are further improvements necessary, and how can they be achieved?

The big unknown of a plain bearing is the material compound itself. Up to now that was seldom questioned. The selection of material was always based on its technical properties. But what is the quality of information about these technical properties?

Again and again bearings will be assessed by running on test facilities, even though the behavior of the material compound is almost unknown. This book focuses on investigating these tests and comes to some interesting conclusions.

Different lining procedures have been developed. Which of them is to be preferred? Are the results comparable? Are the procedures always logically chosen? The techniques of lining after many years remain unchanged. As a result, the further development of lining materials is blocked. What are the prospects for the future?

Often it is good practice to question established "facts." This can allow recognition of new potential, lead to fresh approaches, and eventually present new results. This will be illustrated by the example of development of a new lining process.

Plain bearings are subject to quality control. Is the actual practice of quality control reasonable and sufficient? Are the relevant standards helpful? Are the technical properties of the lining material, as documented in the data sheets, truly representative?

The high complexity of plain bearing technology requires a structured approach to surveys and record keeping. This will be shown in detail with the example of damage analysis. Similar attention is very helpful in other areas of bearing technology.

The foregoing observations and questions, and many more, are addressed in detail in this book, together with suggestions for further approaches to research.

Chapter 2

The Hydrodynamic Plain Bearing and Its Advantages

Hydrodynamic plain bearings—especially compound bearings with tin-based linings—offer many unique advantages over other bearing systems. However, the most outstanding advantage of tin-based lining alloy is often overlooked: it works as a fail-safe component. When any failure causes bearing damage, the tin-based lining will be destroyed but never the sliding partner, the shaft.

A compound plain bearing consists of a backing that is lined with a bearing alloy.

Tin-based bearing alloys work as a predetermined breaking-point safety component.

An outstanding example of this feature is a turbine's bearing. Here a failure resulting in rotor damage is absolutely unacceptable. Therefore only plain bearings with tin-based linings (aka white metals or Babbitts) are used for this application. A further advantage of these materials is that in the event of a bearing problem, failure is gradual. Both characteristics are essential advantages compared with bronzes, which during a bearing failure also damage the shaft, and the AlSn alloy linings that fail spontaneously. A further advantage of all hydrodynamic bearings is the low-noise operation. They give impact- and vibration-damping operation up to the highest sliding speed. Plain bearings are largely tolerant of dynamic impact. Dirt with a particle size below the oil gap width does not affect the bearing. Single bigger particles become safely embedded in the lining. So there is a wide area of tolerance to contamination. Plain bearings need little space and can be designed as a split bearing for easier replacement without removal of the rotor. A little lift of the rotor is sufficient. This advantage is used, for example, on steam turbines. On large vertical machines (e.g., Kaplan-turbines in hydro power stations), the journal bearings consist of up to four parts to minimize the work of assembly (see Fig. 2.1). With other bearing systems, such a comprehensive adaptation of design to the particular requirement of assembly is not possible by any economic means. Another significant advantage is that white-metal-lined plain bearings do not require special materials for the shaft. In the event of a failure, there will be no seizure with the shaft. The possibility to repair plain bearings for reuse makes them very economical.

Plain bearings are normally individually designed for the particular application. This can sometimes appear to be a disadvantage because the designer cannot select the bearing from a catalog. If in this situation there is no competent advice available, very often the designers switch to alternative bearing systems without knowledge of the disadvantages that can result from the decision. Wind-energy systems provide a good example of this. Their roller bearings can be a cause of rotor breakage, as already has been proven in practice. Also, each change of these bearings demands complete disassembly of rotor and propeller. Nevertheless, the makers retain this design, obviously because they fail to recognize or they choose to ignore the interaction between design and typical problems.

Fig. 2.1 Journal bearing of a Kaplan vertical turbine with a shaft diameter of 625 mm. The cylindrical design bearing is divided into four parts to allow economic assembly in a cramped location (1).

A further advantage of plain bearings is the long service life. It is assumed as infinite, and the bearing calculation concentrates on the reliability, taking account of the variables. Safety is assured when the bearing runs at calculated temperature and when wear is zero. The lining material is not included in the calculation. Up to now this approach was sufficient and successful. But this is valid only as long as the load level is below the fatigue capacity of the material compound. About these correlations we have insufficient knowledge. In the future, when plain bearings are required to carry higher load, the fatigue capacity must be given serious consideration to assure long service life. These interactions will be often referred to in this book.

Ultimately there remains the question of what the disadvantages of plain bearings may be. Here we must mention the applications with low sliding speed in which no hydrodynamic operation is possible.

In such cases, self-lubricating sintered bearings, bearing materials with high graphite content, or roller bearings are preferred.

2.1 The Operation of Hydrodynamic Plain Bearings

This book will avoid the details of calculation of plain bearings, which have been repeatedly covered since Vogelpohl [1] and Lang, Steinhilper [2] by many competent colleagues. The experience from practice shows clearly that failure of bearings very seldom can be traced back to calculation faults. Predominantly responsible for bearing failure are misinterpretations of operational conditions and consequent wrong assumptions.

To understand the operation is the key to successful use.

For the successful operation of a bearing, lubrication is crucial. This basic fact had already become clear 4000 years ago. The Egyptians used liquids to reduce friction and wear on the sliding carriages used for transportation of their massive statues. That is shown on reliefs from 1880 BC. The radial bearing gained importance with the invention of the wheel. In 1883 AD, Tower detected the formation of hydrodynamic pressure when he made experiments with railway bearings.

Hydrodynamic pressure occurs as soon as a convergent lubrication gap is created, and the lubricant is transported through the gap by the relative speed of the sliding partners. There are various types of hydrodynamic bearings, but they all operate on the same basic conditions. They all are tribological systems.

Definition according to DIN 50323: Tribology is the science and technique of surfaces in interaction under relative speed. It covers friction, wear, and lubrication and includes boundary surface interactions between solid bodies or between solid bodies and lubricants or gases.

Hydrodynamic bearings operate on an ingenious principle that allows a contact-free and wear-free power transmission from one sliding partner to the other one. Preconditions are the presence of a convergent lubrication gap, the presence of a lubricant, and movement.

2.1.1 Journal Bearings

On a hydrodynamic journal bearing, the convergent lubrication gap occurs by the eccentric location of the shaft in the bearing bore. Here the eccentricity is based on the fact that the shaft diameter necessarily is a little smaller than the diameter of the bore.

The hydrodynamic pressure created by shaft rotation is parabolically distributed, and the maximum is in the rotation direction directly in front of the point of minimum oil gap. In the axial direction, there is also a parabolic pressure distribution. That means that toward the axial bearing ends, the pressure drops.

As soon as there is balance between the acting load and the developed pressure, the shaft is lifted up on the oil film, and contact-free, wear-free hydrodynamic operation exists, as shown in Fig. 2.2. The lubrication gap created is "h." For operational safety, there is a minimum distance h_{min}.

Edge running is a typical problem of radial bearings. For a symmetrical parabolic load distribution along the cylindrical length of a bearing, the shaft and bearing bore must be parallel. That is the theory. In practice, full alignment can never be

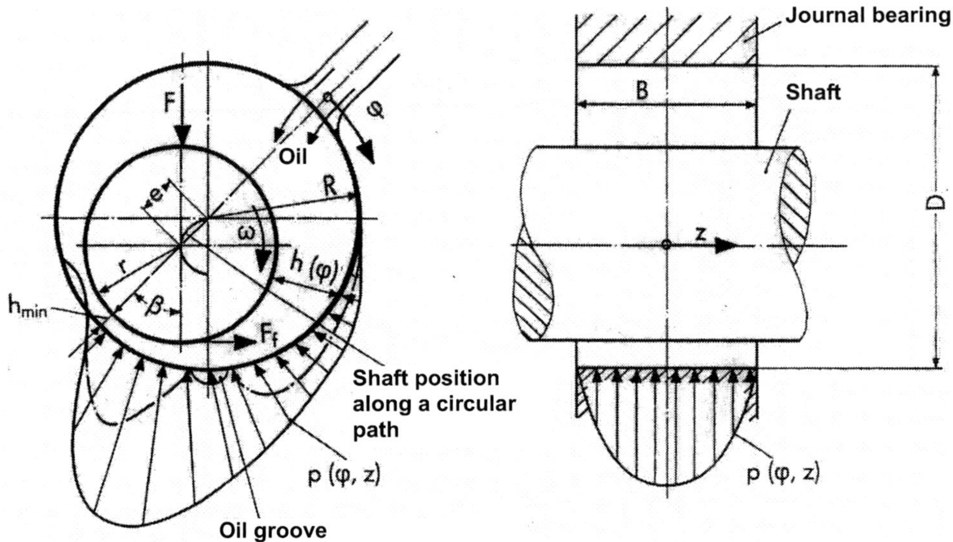

Fig. 2.2 Operational principle of a journal bearing.

guaranteed. With any deviation, the lubrication gap along the bearing length will differ by increasing at one end and reducing at the other. That is the start of problems. First, there is the fact that the real conditions differ from conditions assumed during calculation. The maximum load shifts from the center to the end and increases at the same time as the lubrication gap decreases. It would be convenient if the lubrication compression would increase in the decreased gap to compensate—but, no chance! The lubricant follows the way of least resistance toward the wider gap. That causes a lubrication deficit in the area of minimal lubrication gap and maximum load. Contact between bearing and shaft follows with mixed friction, wear, and eventually damage.

Mixed lubrication is the condition where the oil gap is so far reduced that contact-free sliding between bearing and shaft is no longer guaranteed.

When a case of edge running occurs, first the exact circumstances should be noted in detail. Based on this, a systematic approach must be developed. When mixed friction happens, the preferred approach will always be to try to improve the conditions by the addition of thin layers of material with better sliding properties. In this way, friction wear can be reduced—but for how long? The excellent conformability behavior of tin-based layers will be pointed out when edge running is a problem. At least with tin alloys no damage is done to the shaft. But all these actions and features do not prevent the wear on the bearing, nor do they improve lubrication conditions in the critical area. They are not the ideal approach to solving the problem.

Simply trying to influence the visible results of a problem is always the worst approach.

Most importantly, the reason for a problem must be identified and then conditions must be developed that help to avoid a recurrence. This correlation will be detailed in Chapter 12, "Plain Bearing Damage."

The described case involved a deficit of lubrication local to the bearing edge. Such a lubrication deficit cannot be compensated by modification of the layer: not with a change of composition, nor with an additional layer.

The reason for the problem of locally insufficient lubrication is the lack of alignment between bearing and shaft. A successful approach must aim to achieve correct alignment under all operational conditions. This is the only chance to solve the problem. Granted, often it is extremely difficult to realize; however, it is the only correct approach.

Problem solving means to identify the reason and not just to treat the symptoms!

When conducting an investigation, the lubrication can never be neglected. In most cases the assumption will be made that sufficient lubricant is always available at the right place. That is a big mistake.

A precondition for safe operation of a hydrodynamic bearing is a specified or automatically adjusted convergent lubrication gap to build up a stable lubricant pressure.

The most common misinterpretation involves the assumption that sufficient lubrication is available in the right place.

The conventional radial bearing is cylindrical. With a change of rotation speed and load, the position of the shaft will be changed along a circular path, the so called Gümbel-semi-circle, illustrated in Fig. 2.2. With increasing rotation speed and with decreasing load, the shaft position moves closer to the center of the bearing bore. Under such circumstances, the necessary eccentricity for hydrodynamic operation is no longer present. Consequently, the bearing falls into an unstable running condition and damage follows. Readers that are familiar with bearing technology will think that this is no news! But please remember that when problems occur, there are always conditions that are not being fulfilled, including well-known conditions. It is recommended, therefore, to first check the familiar operational parameters to establish whether or not they are fulfilled.

What is a satisfactory cure for instability of a running bearing? To reduce the bearing play is not a good idea, first because the general geometry, hydrodynamic pressure, and location of the shaft will not be significantly influenced. Second, and more serious, the flow-rate of lubricant becomes reduced. A bearing always needs a sufficient amount of lubricant to carry away heat generated by liquid friction. If the amount of lubricant is reduced, the heat balance is lost and the bearing becomes overheated.

An undesirable dislocation of the shaft can be solved with a suitable bearing design (e.g., the lobed bearing).

The first step should be to check whether all preconditions for proper bearing operation are fulfilled!

2.1.2 Lobed Bearings

The relocation of the shaft in a radial bearing, mentioned earlier, is greatest when sliding speed is high and the load is relatively low. This is frequently the case under power-down conditions, and there are also many applications in which the normal operational state is similarly affected.

The simplest solution is the lobed bearing, double bored to give two lubricant wedges (lemon bearing). It is a journal bearing in two halves, where the clearance at the joint line is bigger than the clearance in the direction of the load. Manufacture is easy to achieve by first machining a cylindrical bore, then machining the extra side clearance either by taking away existing shim plates from the joint faces or by additional machining of the joint faces. The result is a lemon-shape bore, giving one lubricant wedge in the load direction and one opposite. For turbine bearings, it is a straightforward solution to avoid unstable running while shutting down the machine. The radial load will already be reduced while the sliding velocity is still high.

Bearings operating at very high speeds need the multi-lobed design for normal operation. Multiple circumferentially distributed lubricant wedges stabilize and centralize the shaft. Lobed bearings are available in fixed designs, as shown in Fig. 2.3. Very popular is the tilting-pad design of journal bearings (see Figs. 2.4 and 2.5). These adapt optimally to all operating conditions up to extremely high sliding velocity. The bearings are then very highly thermally loaded.

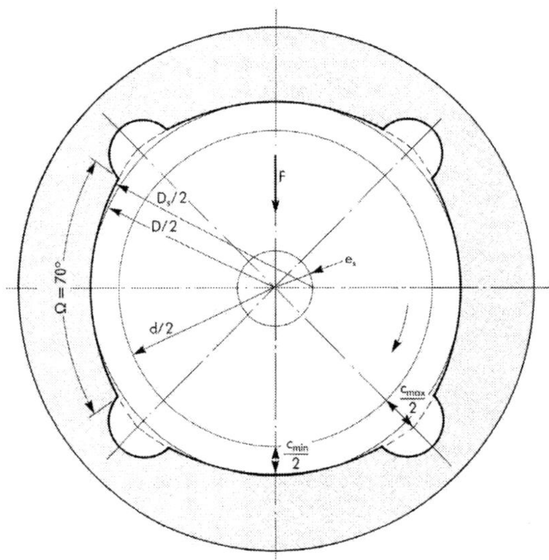

Fig. 2.3 Lobed bearing with four fixed surfaces.

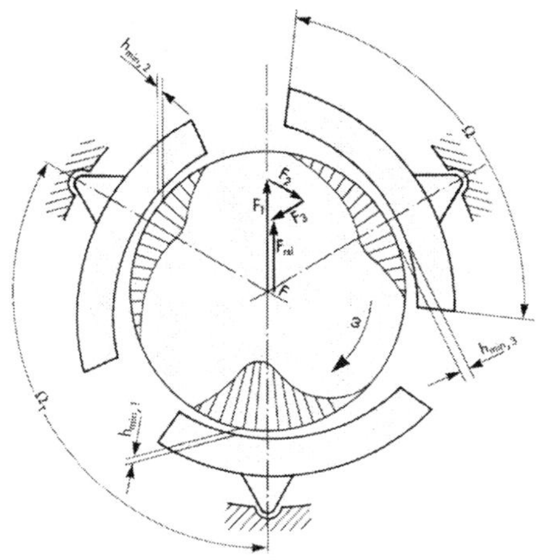

Fig. 2.4 *Tilting-pad journal bearing.*

Fig. 2.5 *Example of a tilting-pad journal bearing (2).*

2.1.3 Thrust Bearings

For transferring loads axial to the shaft, thrust bearings will be used. The load along the shaft will be transferred via a thrust collar to the thrust bearing. The thrust bearing can be designed as a stiff ring with wedge faces alternating with faces parallel to the thrust collar. Very popular is the design with tilting pads. Figure 2.6 shows center-supported tilting pads. This design can be used for rotation in both directions. Figure 2.7 illustrates a thrust bearing with tilting pads; Fig. 2.8 shows the parts of a thrust bearing.

In principle, the operation of a thrust bearing is identical with the operation of a journal bearing. But some differences have to be considered. With the tilting-pad thrust bearing, there is no oil wedge gap before the machine starts. All pads are in contact and parallel to the thrust collar. An adequate design must guarantee that immediately when the machine starts, lubricant comes between thrust collar and pads and forms the necessary wedge gap. This can be achieved by machining the leading edge of each pad with a radius. As soon as motion starts, a lubricant pressure will be built up at the radius along the edge. It lifts the front side of the pad and, at the same time, lubricant enters between thrust collar and pad. The pad tilts and the wedge gap between thrust face and pad is formed.

There are often discussions as to whether the radius along the front edge of the pad is sufficient, or whether it should be extended by a wedge-shaped lubrication inlet behind the radius. As soon as problems occur with the pads, it seems natural to add this wedge or to extend an existing one. Indeed this will make the lubricant inlet easier, but at the same time, the plane surface necessary for operation of the bearing will be reduced. This is a significant disadvantage and overall represents a loss of capacity for the pad.

When designing improvements, there must be verification that disadvantages do not simultaneously develop to negate or reverse the positive effect.

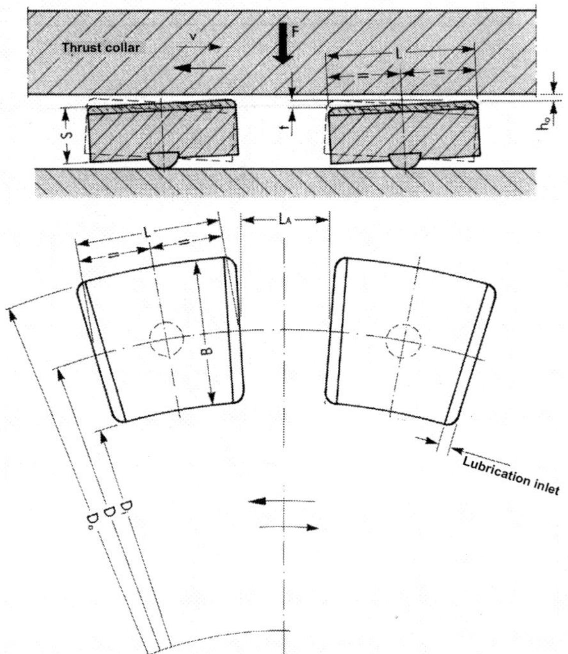

Fig. 2.6 *Principle of a thrust bearing. Example with center-supported pads for use in both rotation directions.*

It is assumed that the necessary radius along the front edge of the pads is present. Nevertheless, when problems arise, full attention should be concentrated on the supplied lubricant, its quantity, the distribution, and the temperature. In general, a bearing needs a supply of lubricant at a defined temperature. On thrust bearings, the feed to the pads is by lubricant injection at the thrust collar in front of each pad. If this is not correct, the lubricant transfer fails. The oil will travel from one pad to the next and the temperature will rise out of control. Oil wipers behind each pad cannot solve the problem. On the contrary, they can lead to turbulence and foam formation.

Thrust bearings have a special problem with centrifugal forces. The lubricant will be thrown to the outside of the rotating thrust collar. The actual supply to the pad, therefore, differs from that planned. Another problem is that pads suffer deformation during operation from mechanical, and particularly from thermal, influences.

This means the lubrication gap increases toward the pad periphery to a significant extent. The consequence is a reduced lubrication gap at the center of the pad and an increase in the maximum load p_{max}. The associated positive effect is an increased amount of cooling oil for the pad and a decrease in the operating temperature. This, of course, depends on sufficient oil flow to completely fill up the increased oil gap.

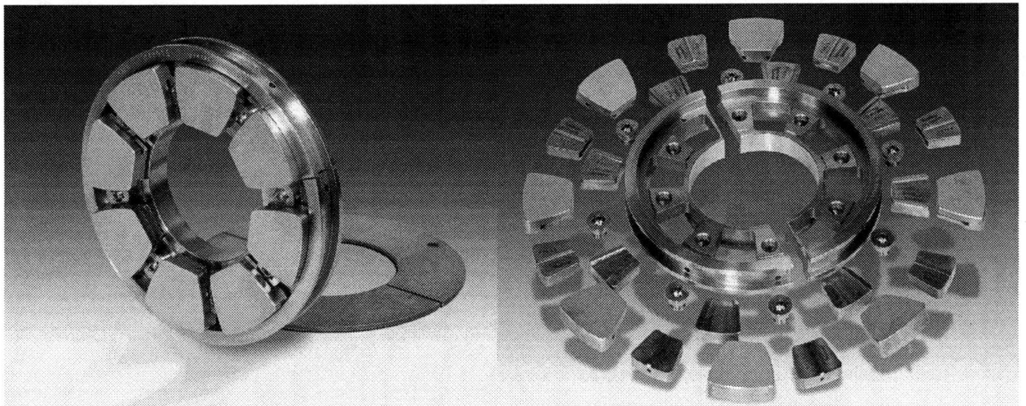

Fig. 2.7 Thrust bearing with tilting pads (2).

Fig. 2.8 Individual parts of a thrust bearing (2).

If the deformation is not considered by the design calculation program, the actual minimum clearance gap will be smaller than calculated and approaches a state of mixed lubrication. If the deformation is asymmetric or the oil supply is not sufficient for the complete pad surface with the increased gap width, then the area taking the load is reduced and the actual load distribution differs from that planned. If a tilt of the pads across to the sliding direction is detected, this should not lead to general surprise.

A systematic investigation of the circumstances becomes essential. To reduce the mentioned problems, a compact geometry of the pads is ideal, and the same applies to tilting-pad journal bearings. There are already tilting-pad journal bearings and tilting-pad thrust bearings available with circular pads. They are well-proven—see Figs. 2.9 and 2.10—but unfortunately this design is not widely distributed, although it suits the ideal.

Thermal influences as expansion or bending can change the basic conditions of a plain bearing.

Problems always arise when real conditions differ from the assumptions.

Close attention should be given to the rigidity. It needs to have sufficient dimensions of the thrust collar to avoid operational load causing unacceptable and unequal deformation. A split and bolted thrust collar cannot fulfill the conditions. Its irregular stiffness distribution leads to a ripple surface and consequently bearing damage. It is extremely important to machine the thrust collar flat and true. This is of higher priority than the surface roughness.

The uniformity of load distribution to all pads of a thrust bearing is only achieved when all pads are located at the same height above datum. Therefore, accurate assembly is essential. Also important is the uniformity and stiffness of the base plate that supports the thrust pads. It must allow no unequal deformation. If, for example, the frame of a big machine has an access hatch

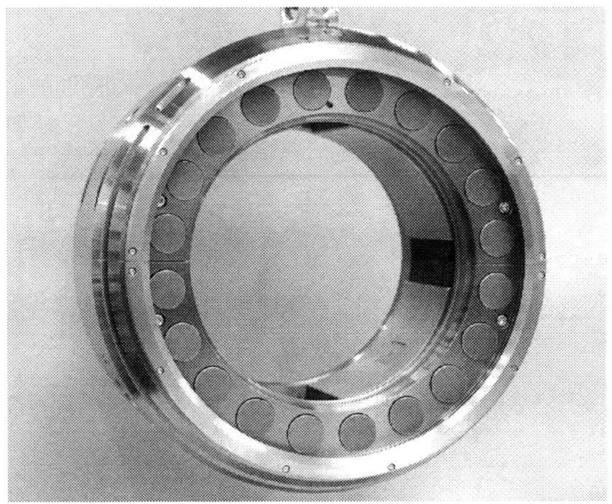

Fig. 2.9 Combined axial-radial bearing with tilting circular pads on the thrust bearing (3).

Fig. 2.10 Journal bearing with circular tilting pads (4).

underneath the thrust bearing, then the pad above this manhole is more elastically supported than the others. Its load becomes reduced and the other pads have to carry more.

If a loaded base plate deforms radially to the outside, then pads have to follow this deviation when they sit on a linear pivot. Parallelism of the lubrication gap is no longer possible, and the pads get mixed friction along the inner diameter.

When designing thrust bearings, the stiffness of the surrounding (bearing base plate, thrust collar) is crucial.

For thrust bearings, R & D is needed regarding the interaction of load, speed, temperature, geometry, mechanical and thermal deformation, sufficiency of lubricant supply to fill lubrication gap, and pressure distribution.

2.1.4 Hydrostatically Lubricated Bearings

For hydrodynamic operation free from metal-to-metal contact, minimum requirements of rotational speed vs. load parameters have to be fulfilled. If the speed remains too low or the machine is at low speed with high load, then the bearing remains for an unacceptably long time in mixed friction conditions.

For such operating conditions, a hydrostatic lubrication system is recommended. High-pressure oil will be fed from outside pumps (approx. 6×) directly to the area of main load. The high-pressure oil avoids the contact between sliding surfaces.

When bearing damage occurs during start up of a machine, despite the use of a hydrostatic lubrication system, we get with wonderment the confirmation that oil pressure was fully there. But with this information comes no confirmation that the oil actually reached the bearing! After repair work it often happens that the high-pressure oil system is blocked. The reason may be residues that came into the system during repair work, but a previous bearing damage may also be responsible. This happens when during the course of damage the white metal melts and flows backward into the high-pressure oil pipe to block it.

Also important is the complete venting of the high-pressure system before commissioning the machine. All the mentioned problems can be detected or solved by switching on the hydrostatic lubrication when the machine is still at a standstill. At the same time, the lifting up of the shaft has to be checked with a test gauge.

For hydrostatic operation, the diagnosis of correct pressure is not sufficient. Oil feed must also be sufficient!

Usually the hydrostatic lubrication will be switched off as soon as the conditions assure hydrodynamic operation. For proper hydrodynamic operation, it is important to lock the hydrostatic system with a non-return valve. Otherwise the hydrodynamic pressure can be lost through the hydrostatic system, and damage is then unavoidable.

Although the mentioned problems lead to failure of the hydrodynamic operation, operators repeatedly favor working on the oil pocket of the hydrostatic system. This is a typical example of dealing with symptoms based on a misinterpretation of the correlations.

It should be strictly questioned whether what we do, and the reasons we give to justify our actions, provide the solution.

Hydrostatically supported bearings are sometimes equipped with very complex designs for the area of hydrostatic oil inlet. In the past it was to some extent justified. Cast iron backings have low bond strength with the bearing layer material, and therein lies a risk of penetration of the high pressure oil between backing and layer. To avoid this, some very extensive and somewhat excessive designs have been developed. A simple bronze screw plug assembled at the end of the tinning process on the backing is absolutely sufficient to solve the problem.

For backings made from steel, as is now state-of-the-art, good bond is always achievable, and therefore no special provision for the hydrostatic hole is necessary. This is an example of how approved design loses validity because of changed conditions.

Lubrication grooves should not be located in the area of hydrodynamic pressure because they disrupt the oil gap geometry and the formation of hydrodynamic pressure. For the same reason, the lubricant groove for the inlet of hydrostatic oil should be designed as small as possible. The oil inlet for hydrodynamic operation of journal bearings should be in the unloaded area. There, the pressure is low and oil is freely admitted. For bearings carrying a load of constant direction, it is much easier to fulfill these rules than for bearings for which rotating load applies.

Chapter 3

Backing Materials

3.1 Steel Backings

The ideal material for the backings of compound plain bearings is steel; forged steel is preferred, although cast steel is often chosen for complex backing shapes for convenience of production. Structural steel can also be used. Typical steel qualities for compound bearings are: C10, C15 according to DIN 17210, GE 200 (formerly GS38), GE240 (formerly GS45) according to DIN EN 10293, S235JRG2 (formerly RSt37-2), and S355J2G3 (formerly St52-3) according to EN 10025.

Good tinning is the essential precondition for good bond between backing and lining material when using soldering or casting processes. Good bond is assured if the formation of the intermetallic compound $FeSn_2$ is possible to a sufficient extent during the tinning process. The chemical composition and the microstructure of the backing material both have considerable influence on the bond quality.

The microstructure of steel with a low carbon content is characterized by solidification according to the "metastable system" and results in ferrite + pearlite + cementite. Pearlite is the eutectic mixture of Fe + Fe_3C (ferrite + cementite), and is not involved in the bond. The tin can form a compound only with the ferrite crystals of the backing material. The higher the proportion of carbon, the more pearlite there is in the microstructure and the lower the bond strength.

For optimum bond, a carbon content of < 0.2% is desirable. Steel with a higher content of carbon should only be chosen if reduced bond quality is acceptable. Also harmful to bond quality are chromium above 0.25% and nickel above 0.5%. These influences have been noted especially in cast steel. Unfortunately, values for manganese >0.4% and silicon >0.25% can also lead to bond problems. Obviously the composition is not alone in affecting the bond quality. The homogeneity of the backing material is also influential.

The suitability of a backing material for achieving good bond has been decided empirically over the last hundred years. Unfortunately, the constitutions of many steel grades have been modified, especially in recent years due to the international harmonization of standards, and we now have the situation that problems can arise with producing a good compound bearing while using supposedly well-proven materials.

For selecting backing material, the advice of the bearing producers should be taken into account.

Best-quality material still is unsuitable if no compound plain bearings can be produced from it.

3.2 Cast Iron Backings

In former times, cast iron was preferred when the backing geometry had a complex shape. The good damping behavior was highlighted as an advantage. In practice, the damping was simply due to the typically large mass of old cast iron designs. The mass has a dominant influence on damping compared with any variation in material specification.

The essential disadvantage of cast iron is the poor quality of bond with the lining material. Nevertheless, cast iron has been used for a long time, and efforts have been made to compensate for the weakness of bond by design provisions. A mechanical locking design, known as dovetail grooves, actually makes the quality situation even worse. Chapter 8 gives more information about those dovetails.

A mechanical joint between steel backing and layer material is not state-of-the-art for modern and highly loaded plain bearings.

What is the reason for low bond with cast iron?

Cast iron used for backings has a carbon content of 2 – 4.3%. On a stable solidification, the microstructure of cast iron consists of ferrite + graphite.

This ideal condition requires solidification with an extremely slow cooling rate. In practice, such cooling conditions cannot be realized, especially not when the backing has big variations of wall thickness. Therefore, in practice the solidification follows the metastable system. That means that beside the graphite, pearlite will also be formed. Thus, the microstructure composition is ferrite + pearlite + graphite. As mentioned previously in connection with steel, the compound bond quality drops with increasing pearlite. The graphite in cast iron also hinders the bond by obstructing the wetting of the backing with the molten white metal.

To achieve an acceptable bond, the following limits should be observed:

Si<2.5%

P<1.2%

C<3.35%

If the graphite is smeared across the backing surface during machining, then practically no bond can be obtained with the cast lining material, so before tinning the cast iron backing, the surface will be treated to reduce the graphite content.

Expensive procedures such as the direct-chloride process and the Kolene-E procedure are no longer used due to the high costs and the decreasing use of cast iron for backings.

Bond quality equal to that of steel backings is not achievable with cast iron.

3.3 Copper Alloy Backings

Copper alloys are used for solid plain bearings and for backings of compound bearings (i.e., those lined with bearing alloy).

When comparing the technical properties of the bronzes with tin-based white metals, bronze seems to have a clear advantage. Tin-bronze can cope with 10× the load and operate at up to 400°C for bearing applications.

At first glance, these advantages may be assumed to be conclusive. But practice is different, and the application of bronzes for plain bearings is restricted. To use a bearing at 400°C requires special lubricants. They have to provide sufficient viscosity at this temperature and may age in a short time and need replacement. Such lubricants are not liquid at room temperature, and therefore special arrangements are needed to ensure adequate lubricant supply when the machine is starting cold.

Copper alloys used for bearing running surface applications bring with their high matrix hardness the risk of seizure and damage to the shaft. Hot cracks in the steel shaft may penetrate some millimeters deep. Such damage to the shaft is, for most applications, not acceptable. The problem can be eliminated by selection of a hardened and tempered shaft. With tin alloys, such problems are unknown and no hardening of the shaft is necessary.

Using copper alloys for solid plain bearings, or as lined backings, the thermal expansion coefficient has to be considered. For tin alloys, the coefficient is about $22 \times 10^{-6} \times K^{-1}$. Bronzes have a coefficient of about 17.3 and C15 steel about 11. For steel or iron bearings lined with a tin-alloy layer, the different coefficients are only likely to show effects during fabrication; however, for bronze bearings there can be influences on the functionality of the bearing.

If a radial bearing has a quick start from cold to operational conditions, there is initially a big temperature gradient between the bearing surface and the housing of the machine that is still cold. In this situation, the thick-walled bronze, with its considerably bigger thermal expansion coefficient, is restricted in its expansion. The consequence is a loss of running clearance during start up. If this temporary condition of reduced clearance has not been considered in the design calculation, overheating of the bearing follows due to reduced lubricant flow rate. In extreme cases, the plain bearing operates as a brake.

Copper alloys, especially copper-chromium alloys, are popularly selected as backing material for the pads of thrust bearings. The motivation is always the good heat conductivity of this material. This at first seems to be a correct decision; the more so as practical experience seems to provide support. But do the results really confirm the assumption? Plausibility as a justification is surely not sufficient. To avoid mistakes when using backings with copper alloys, the precise circumstances have to be assessed and considered.

Good heat conductivity of a material only becomes effective when a good heat transfer can be guaranteed. That means that the heat flow from the back of the pad must not be hindered. But with pads, this is exactly the problem. On the back of each pad there is just a small area in contact with the base plate, which is usually steel. Apart from that there is only contact with the lubricant. For both of these contacts the heat flow is limited as much as it would be with steel pads. Thus, improved heat conductivity cannot be the explanation for the lower temperature peak on copper pads.

In fact, the higher temperature expansion coefficient of copper compared with steel, in combination with the temperature gradient, produces significantly more bending deformation of the pad. This increased bending deformation leads to an increased lubrication gap at the pad edges, and as a result the pad gets more cooling oil. Furthermore, here the conditions again apply, as mentioned in Section 2.1.3 for thrust bearings from steel. It seems worthwhile to consider whether the use of copper alloys is essential and adequate for the application in question.

A good result does not exclude the need to know how it came about.

3.4 Heat Treatment of the Backings

Practical experience with cast bearing backs consistently shows that heat treatment is essential for good bond between backing and lining material. For compound bearings, three different heat treatments are relevant: normalization, stress-relief, and hydrogen removal treatment.

For the following treatments, it should be noted that for the recommended hold time, the time needed to heat right through the casting has to be added. It depends on the power and capacity of the oven and on the wall thickness of the backings.

3.4.1 Normalization

This heat treatment is in use on steel backings to reduce the stresses and simultaneously regenerate and refine the microstructure for improved material properties and ductility. Stresses from steel production by casting and hot- or cold-forming must be minimized, so for new backings the normalization treatment is essential.

In the case of repairs to plain bearings, there is no need to repeat this treatment unless the bearing backing has experienced a change of microstructure. This can happen during welding work on the backing.

The annealing temperature depends on the carbon content and is 30–50°C above the G-S-K line on the FeC diagram, shown in Fig. 3.1. For steel with carbon content <0.2%, the temperature is about 910°C. The exact temperature for the particular material should be taken from its material data sheet. The material should be heated up as quickly as possible, and the hold time should not be excessively extended. A rough indication is one minute per millimeter wall thickness. After the treatment, slow cooling is needed to avoid thermal stresses in the material. It is important to avoid overheating and excessively extended treatment. In practice, the previous treatment is frequently mixed up with the next one. New backings that have not been given a normalization treatment very often show bond failure.

Omitted or improper heat treatment of the backing leads to bond failure. New backings always need normalization treatment.

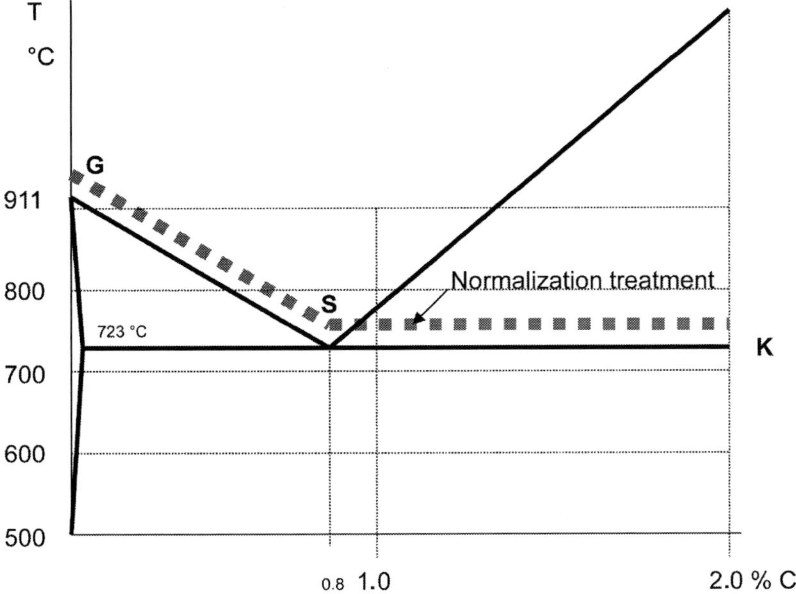

Fig. 3.1 FeC diagram with the area for normalization treatment, 50°C above G-S-K-line.

3.4.2 Stress-Relief Treatment

Stress-relief treatment is applied to reduce stresses in the backing resulting from machining and preparation work. This treatment should be made at a temperature of 530–550°C and a hold time of about two minutes per millimeter wall thickness; see Fig. 3.2. After the treatment, slow cooling down is needed. There will be no modification of microstructure, which has already been refined with the normalizing process. Hence, the stress-relief treatment will only be needed when repairing plain bearings.

Plain bearing repair should always be done in combination with stress-relief treatment.

3.4.3 Hydrogen Removal Treatment

Production steel usually has a hydrogen content of about 5 ppm. The hydrogen is completely in solution and diffuses out slowly. Depending on the material

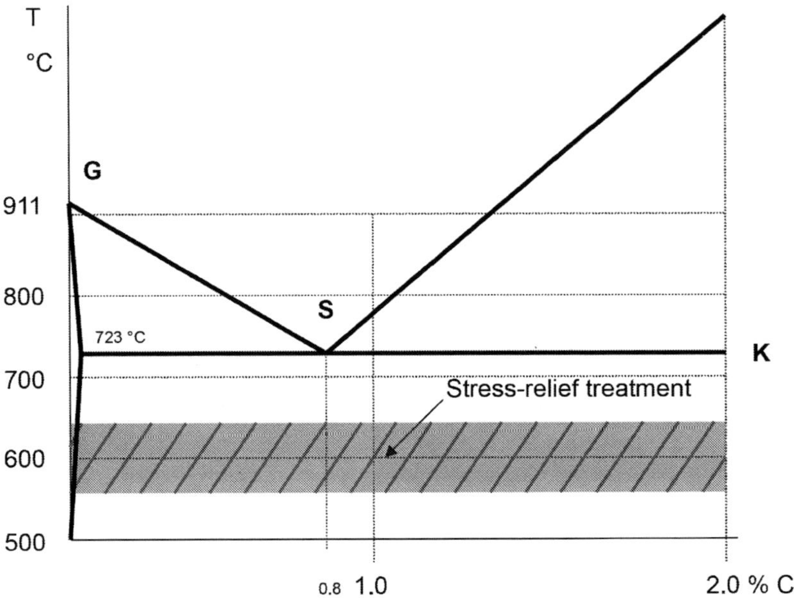

Fig. 3.2 *FeC diagram with the area for stress-relief treatment about 600°C.*

thickness, after four to six months of normal storage, the material is largely free from hydrogen. Early rough cutting of the material increases the surface and speeds up the diffusion process. For general use in industry, there is no problem with the hydrogen because the diffusion is not hindered. On compound bearings the lining layer is a barrier against diffusion, and for hydrogen content above 1.5 ppm the damage risk is serious. The smallest imperfection in the bond allows the hydrogen to change from atomic to molecular condition, the volume increases dramatically, and the white metal lining becomes separated from the backing. A failure with characteristic blisters of the lining occurs (see Figs. 3.3 and 3.4). Often it is assumed that the problem can be avoided by ordering vacuum-treated steel with a certificate which confirms the hydrogen-free condition after treatment.

Consider the following: the material test piece will be taken from the liquid charge, it is analyzed, and the result confirms it is hydrogen free. The problem is that the liquid steel charge again absorbs hydrogen during transport and casting. The hydrogen comes from the atmosphere and from any moisture in the brickwork of the transfer ladle. Before solidification, the steel has again absorbed quantities of hydrogen, in most cases well above the critical value for plain bearing linings of 1.5 ppm.

> *The certificate of vacuum treatment to a level of zero hydrogen in the steel is not a guarantee of sufficient hydrogen reduction for use as backings of plain bearings.*

A further problem is the near impossibility of establishing the hydrogen content of the bearing backing. The hydrogen content is located on the center of mass. To get a specimen for analysis from there, the backing has to be destroyed

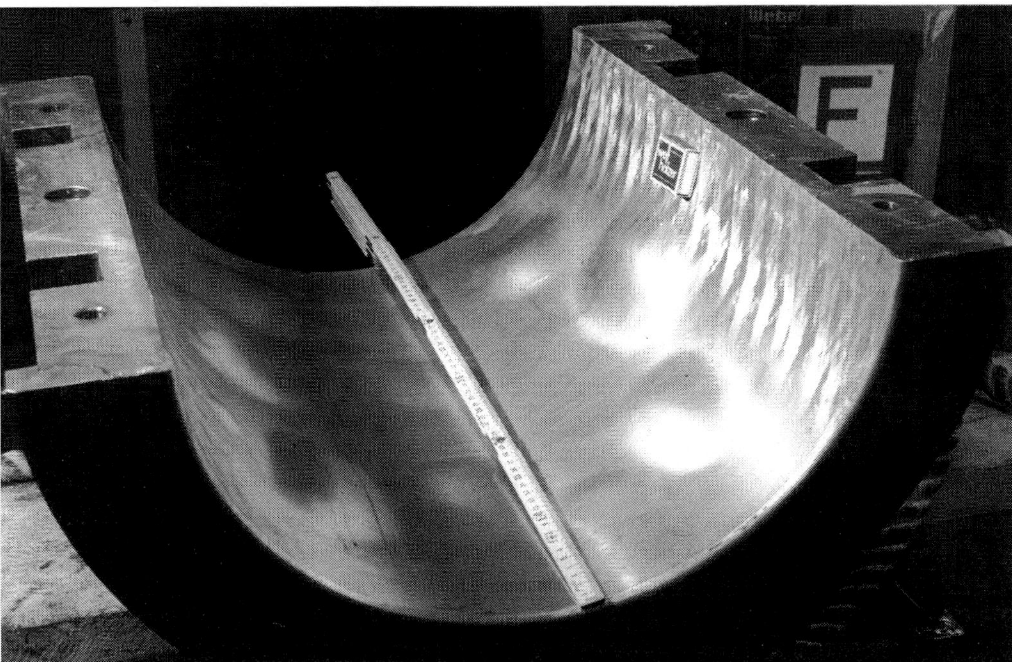

Fig. 3.3 *Example of damaged lining due to hydrogen-filled blisters.*

completely. Even if this were undertaken, the exercise defeats itself. Surfaces are increased during the machining, especially when the specimen has a big surface in relation to its mass, and there is heat generated during cutting. These factors aid diffusion, and in practice the hydrogen has already gone from the specimen by the time it is ready for tests.

> *In practice the real hydrogen content of a bearing backing is not detectable.*

How to avoid problems regarding the hydrogen phenomena? The flow of hydrogen through the material is a function of distance, time, and temperature. The ideal way to speed up the diffusion is a heat treatment with as high a temperature as possible. But there are unacceptable side effects. The ideal high temperature for quick hydrogen diffusion ruins the steel. A breakdown of the cementite, forming methane, can result in decarbonization and brittleness of the steel.

> *With each action, consider the consequent secondary effects.*

> *Heat treatment at excessively high temperature will ruin the steel.*

Prolonged heat treatment needs a temperature limit of 500°C to avoid harmful effects on the steel. For hydrogen reduction, therefore, 500°C has to be used with a minimum holding time of 30 minutes per millimeter of thickness. This is the recommended treatment for backing thickness above 60 mm. For smaller thicknesses, the hydrogen treatment is covered already by the normalization treatment. The recommended time-cycle results in steel backings with a large

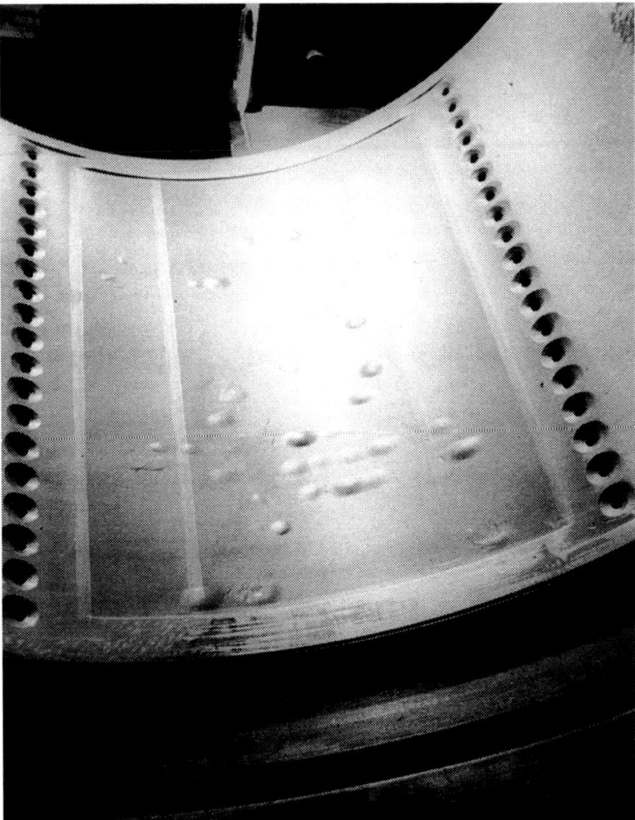

Fig. 3.4 *Example of damaged lining due to hydrogen-filled blisters.*

wall thickness requiring an extremely long holding time. When holding times are 100 or more, there is always a big temptation to shorten the procedure. We expressly advise against that. In practice, such saving actions fail regularly. Hydrogen damage mostly occurs after finish machining, when it becomes very expensive with the heat treatment, lining, and machining all having to be completely repeated.

Consider the risk of hydrogen damage for backings of new bearings with thickness of more than 60 mm.

Chapter 4

Lining Materials

4.1 Lining Materials of the Past

Plain bearings have been in use since the beginnings of industrialization and have undergone continuous improvement. The predominant design now in use is the compound of a structural backing, usually steel, lined with a specialized bearing alloy. The lining is known in the UK as white metal and in the U.S. as Babbitt metal after the inventor Isaac Babbitt, who in 1839 devised an alloy particularly suitable for plain bearings.

The relatively soft bearing lining has to withstand the load under all operational conditions without harming the shaft during the frictional contact of start-up and stopping. The layer has to be as soft as possible and as strong as necessary. These distinctly opposed requirements can be readily satisfied with tin-based alloys.

Hundreds of different alloy specifications for lining plain bearings have been developed over the years, at first lead-based and later tin-based.

Worldwide there are more than two hundred white metals existing today, but reference to the actual DIN and ISO standards provides just a small selection. One reason is the elimination of alloys containing cadmium from the standards under environmental rules. Soon the alloys containing lead will follow, both for environmental and technical reasons. They are of low performance and can be replaced easily by tin-based alloys, which are a more capable alternative. The same applies to tin-based alloys containing lead, such as $SnSb_{12}Cu_6Pb$. In the end there would remain just one standard alloy, $SnSb_8Cu_4$. Unfortunately, this material has modest properties because it tends to creep severely. Worldwide this alloy is standardized under DIN ISO 4381, in Italy under the national standard UNI 4515 with the name MB 90 and in the U.S. under the name ASTM B23 Grade 2. This is just one example of an alloy that exists under different identifications. Taking into account all such multiple listings, the apparently large number of lining materials for plain bearings is actually less than 20 worldwide. Of this small quantity, a number differ only in the specification tolerances and actually have the same technical properties. Ultimately, we find there are only a few standardized plain bearing lining materials available.

4.2 Lining Materials of the Present

In addition to the standard alloys, especially for high-duty bearings, lining materials have been developed to carry higher load, have reduced creep, and achieve longer service life. When developing such enhanced lining materials, a systematic approach is critical, because there are a number of interrelations, interactions, and restrictions that must be considered. The many failures of such development work can be traced back to the fact that the need for interdisciplinary coordination is not recognized or is ignored.

As long as interrelations are not recognized it is impossible to define boundary conditions and to observe them.

Interdisciplinary coordination of R & D is the formula for success.

4.3 Constraints for Tin-Based Plain Bearing Alloys

Using basic material science, it is no difficult matter to develop an improved plain bearing alloy with, for example, a higher tensile strength. Such a new material developed in a laboratory may seem very good, but for practical application a range of conditions must be fulfilled. Metallurgical experience, the alloy manufacture, the techniques of plain bearing production, and details of the bearing's operational conditions all have to be considered.

4.3.1 Viability of Manufacture

First, the new material has to be suitable for economic industrial manufacture.

When a fine microstructure is formed in the laboratory without any problems, it does not automatically mean that industrial manufacture will give equal success. The adding of alloying elements may be handled differently between laboratory and factory conditions. Such deviations may influence the quality. Theory and practice sometimes are a long way from each other. Refining is possible in theory with many elements but in practice often brings risk and adverse reaction.

A typical example will show this:

Refining of white metals can be performed using arsenic, selenium, chromium, MgAl, or silver.

Arsenic (As) goes easily into solution, and a homogeneous distribution is always assured. Arsenic is an established refining element, but it is toxic.

Selenium (Se) is also a refining agent but only in concentrations up to 0.03%. It would be added to the alloy as SbSe. The Selenium content in the SbSe compound varies in practice, and the low Selenium concentration in the alloy needs a tight tolerance range of ± 0.01%. That makes the use of selenium impractical, because the tolerance cannot be achieved in commercial manufacture.

Chromium (Cr) has a good refining effect, but alloying with chromium is in practice problematic. A master alloy has to be produced at high temperature and fed into the bulk alloy. The majority of the chromium goes into slag, and the alloying fraction cannot be adjusted accurately. Although the refining action is similar to that of silver, the process is too complex and the material losses are too high.

MgAl additives have also been investigated. During laboratory tests the fluidity of the liquid alloy was reduced to a level unacceptable for practical applications.

If the MgAl is introduced via a master alloy, the refining effect has already disappeared during the alloy ingoting; however, direct feeding results in the formation of Al_2O_3 crystals. These have a hardness unacceptable in plain bearings because of the risk of damage to the shaft.

Silver (Ag) has a high melting point, 960°C, but it dissolves in tin in small concentrations at 235°C. For this reason it can be processed easily, and a very homogeneous distribution in the alloy is always assured. Silver is, of course, very expensive, but it develops the full refining effect with a concentration of only 0.04–0.15%. This tolerance range can be readily achieved in manufacture of the white metal.

Alloying elements for improving the material strength can bring problems in practice, as the following examples show:

Cadmium (Cd) has been used for over 50 years to improve the material stiffness. Cadmium evaporates from the liquid alloy as the temperature rises and due to its toxicity is increasingly avoided. Cadmium forms a very low melting point eutectic together with lead and tin, resulting in an embrittlement of the material. As long as the alloy is free of lead, this effect can be avoided.

Bismuth (Bi) has a similar effect to cadmium. It forms a very low melting point eutectic together with tin. Because tin is the base matrix of white metals, a eutectic when bismuth is present cannot be avoided. Because the eutectic melts around the level of maximum bearing oil temperature, bismuth in a white metal is absolutely unacceptable. It is defined as an impurity with very low acceptance limits.

Nickel (Ni) has a high melting point of 1452°C and goes into solution in tin only at high temperature and very slowly. To feed Nickel in solute condition to a white metal, a separate master alloy SnNiCu has to be produced at high temperature and introduced as liquid to the white metal.

Lazy alloy producers may avoid this extra work by feeding Ni as powder or pellets to the white metal. The alloy analysis then confirms the presence of nickel in the white metal, and the specification seems to be correct. But in this case it is a mixture and not an alloy, which is not the same thing. Unfortunately, from the analysis certificate the consumer cannot recognize the difference. It is therefore essential to select reputable suppliers that know the original recipe and follow it.

The certificate of analysis does not by itself control the quality. It depends also on compliance with the original recipe.

Zinc (Zn) is similar to cadmium in effect and gives an improvement of material stiffness. There is no tendency to brittleness, such as with cadmium, and the creep is significantly reduced. Zn-hardened white metals are, therefore, much more resistant to fatigue. Zinc assists the formation of slag on the liquid alloy. This is actually helpful, as the slag is a barrier against the atmosphere and avoids continuous oxidation of the melt as well as retaining heat, thus keeping the recommended melting temperature.

The basic elements of the white metals are subject to proportional limits:

Copper (Cu) is, besides tin (Sn) and antimony (Sb), one of the three basic elements of white metals. For a long time, it has been well known that the compressive yield strength increases with increased copper content. With more copper, the melting point of the alloy increases, the fluidity is reduced, and a higher casting temperature becomes necessary. Furthermore, high copper content tends to intensive segregation, and content above 6% leads to a copper seam along the bond interface. This layer is brittle and results in an unacceptably weak bond in compound material bearings.

The quality optimization of the white metal is limited by the casting process.

Finally, impurities should be discussed:

The negative effect of lead (Pb) is consistently underrated. Lead is not homogeneously distributed in the alloy and can segregate. Areas with higher lead concentration form a low melting eutectic of Pb, Sn, and Sb. Conventional material tests (such as tensile and compression) show no abnormality at room temperature. At higher temperatures, the results fall dramatically due to the reduced melting point. Similarly, the impact-bending test gives a clear hint of the negative influence of lead, only at higher temperature. Because this test is not standardized, there are no results documented. Most technical properties are documented for room temperature. That gives the false impression that lead impurity has no adverse effect. For internationally standardized white metals, the lead impurity limit is set at 0.35%. That is clearly too much. Here the international harmonization of standards has been given more consideration than the quality of the alloys, with the result that the goal of the standardization is defeated.

The standards accept high levels of impurity. This leads to a reduction
of technical properties.

It is bizarre that tin-based white metals are in use with up to 3% lead (for example, $SnSb_{12}Cu_6Pb$). This cannot be justified with the argument of good sliding behavior of the lead. Modern plain bearings usually run hydrodynamically; if they go into marginal lubrication for longer than normal, damage is not far away, and the presence of lead cannot help.

The various raw materials used can contain impurities that are, in total, unacceptable. Therefore, for high-quality white metals, high-purity raw materials have to be selected. Furthermore, during the manufacture of plain bearings, there must be no addition of lead impurity such as is caused by using solder for tinning instead of pure tin.

At plain bearing operational temperature, lead impurities will let the technical properties "melt away."

Other impurities such as Fe and Al are limited by the alloy specifications. In practice, these elements will not be found in the raw materials used and, therefore, will give no problems for white metal production.

When selecting elements for alloy manufacture, commercial influences also must be considered. If manufacture with the chosen element is too time-consuming, then costs are increased. Using a very expensive element has the same consequence. Tests with Gold (Au) were very successful from the technical point of view, but the costs are unacceptably high. Further tests with gold have, therefore, been stopped.

4.3.2 Lining Quality

The next step on the way to a good white metal is the ability to be lined onto the bearing backing. During the lining process, there must be homogeneity and no segregation. There must be no tendency during the lining process to lose elements of the alloy by oxidation or evaporation, and the fluidity must remain constant. Finally, it must be possible to obtain sufficient and uniform bond strength between the lining and the backing.

4.3.3 Reliability under Operational Conditions

Even if no problems arise from white metal production through to bearing manufacture, the material still has to satisfy expectations under the operational conditions. As already mentioned, an essential characteristic of a highly loaded compound bearing is that the layer material does not creep. Furthermore, low melting point eutectics are unacceptable. A good illustration is the comparison with steel: white metals operate up to a temperature level of 50% of the melting point. Under similar parameters, steel is already a dark-cherry-red temper color, and use will be restricted. In contrast, the white metals are in long-term service under these conditions.

If the melting point is reduced, it comes closer to operating conditions, and the technical properties are lost. Approaching the liquid stage, there is no mechanical strength.

In view of the importance of this matter, here is a summary of all elements that can cause areas of reduced melting point in white metals: lead forms, together with tin and antimony, a eutectic melting at 183°C. Lead forms, with tin and cadmium, a eutectic melting at 145°C. Bismuth and tin form a eutectic melting at only 139°C. These weaknesses will not be improved by filing a patent with documentation on technical data limited to measurement at room temperature!

Elements which can form a low melting eutectic with the other elements of the white metal should not form part of a white metal. They endanger operational safety.

4.4 The System for Developing a New White Metal

Because of the small number of different existing white metal qualities and the persistent desire for improvement, there have been attempts to develop new lining materials for plain bearings. This has been done almost entirely by the empirical method, for which new elements have been tried, followed by a report on whether an improvement was noted. The improvements reported were, unfortunately, mainly based on toxic elements and, considering the expenditure of time and effort, the result was deplorable. Nevertheless, the available assortment was adequate for a long period of time, especially the white metal TEGO® V738 (5), developed about 1950, which had a high compressive capacity and low creep, both effects thanks to the element cadmium. When a ban of the toxic cadmium became more and more likely, the author felt the need to develop an alternative, environmentally friendly white metal. A quick result was needed and, therefore, to continue with the empiric experiments of the past 100 years seemed to be unpromising. The success, achieved after some months, was the result of fundamental research work to discover the way in which the elements are actually incorporated and the effect. This systematic approach will be presented here.

Further to the previously described conditions for developing a new white metal, the following requirements have to be considered:

The new material must still consist of the basic system Sn-Sb-Cu to avoid a significantly different crystal structure. This is to ensure acceptance and facilitate the introduction of the new material into the market.

The material improvement is based on the following three steps:

1. The alloy has to be made lead free.

2. The composition of the basic system has to be optimized.

3. An element will be added that has been found with the fundamental research work.

Each of the three mentioned steps contributes about the same enhancement to quality. Naturally, this is just a rough indication due to the complexity of the many characteristics and interactions that have to be fulfilled to gain a useful improvement.

4.4.1 Lead-Free

The fact that a considerable fraction of improvement can be achieved solely by a lead-free material shows clearly the importance of avoiding lead impurity in general. This fact cannot be emphasized enough.

As a result of environmental concerns, it can be assumed that alloys containing lead will soon be forbidden. For a long time, the producers of high-grade white metals have used the lead-free option to improve the quality and manufacture to tolerances significantly stricter than those of standard alloys. The permissible lead impurity of these high-grade tin-based white metals is defined as <0.06%.

4.4.2 Optimization of the Basic System

For tin-based white metals, the basic elements are tin, antimony, and copper. It is well known that for each 1% addition of the copper and antimony content, the compressive yield increases by approximately 3 N/mm². Because serious problems can arise during the casting process, an uncontrolled increase of copper and antimony is not feasible. The concentrations are, in practice, limited to 12% for antimony and 6% for copper.

4.4.3 Additional Elements

To find a replacement for the empirically discovered toxic elements arsenic, nickel, and cadmium, used as property enhancers in white metals, is demanding work. We decided for fundamental research work to discover the way in which the elements are actually incorporated and their effects. Once the interaction of the different elements is known, it becomes possible to specifically extend the system and to improve the material.

For arsenic, the research work showed that, in part, it joins the SbSn phase, where it acts as a nucleating agent for refining the SbSn crystals. Considerable quantities will be found within the matrix. It has been decided to replace the arsenic with a low quantity of silver for the desired refining.

Nickel can be detected only in the Cu_6Sn_5 crystals. As all of the nickel used is found exclusively there, any influence of this element on the properties of the material as a whole can be excluded—in particular, the alleged improvement of the sliding properties. This effect of nickel is known from bronzes, but even there only with the presence of several percent. In white metal, a low quantity of nickel does not bring about the desired improvement. For this reason, nickel can be excluded from the alloy altogether.

Cadmium can be detected in about equal shares both in the Sn matrix and in the SbSn crystals. The solubility of cadmium in a tin-based alloy amounts to approximately 1.2% by weight. Only about 0.6% maximum is available for the desired matrix-hardening effect, and it causes a simultaneous, enormous increase of compressive strength and embrittlement of the material.

When developing the new material, the cadmium-containing alloy, TEGO® V738 (5), was used as a reference, and the $\sigma_{d0,2}$ compressive yield of 80 N/mm² at 20°C and 48 N/mm² at 100°C had to be equaled or exceeded.

The impact-bending strength of the reference alloy is, at 250 J, the lowest value of all tin-based alloys. This is a clear indication of the brittleness of the material. Because a lower value is not expected for the research work, this extensive test has been put on hold. The bond test also has been put on hold because no significant deviations were expected. This way, the test procedures after each research step were reduced to a minimum and the working progress improved enormously.

The cadmium replacement has to be environmentally harmless. This is not just a fashionable trend, but is based on economic considerations. In the Scandinavian countries and in the U.S., the import of materials containing cadmium is already restricted. These restrictions will certainly become even tighter in the future. In Europe, the use of cadmium is not forbidden, but the permissible emissions become more and more reduced, and that makes manufacturing plain bearings with alloys containing cadmium increasingly problematic. Also, the disposal problems increase and push up the costs.

The substitute element must be soluble in the tin matrix to achieve a mixed-crystal hardening. It must not form any low-melting eutectic, and manufacturing and recycling has to be efficient. In Table 4.1, all elements soluble in the tin matrix are listed. Aside from gold and zinc, all the remaining elements are toxic, form low melting eutectics, or are difficult to handle. Gold as an alloying element is very good for the technical properties but is too expensive for technical use. So only zinc satisfies all requirements imposed on an adequate substitute for cadmium in tin-based white metals.

In the past, it was assumed that zinc could not be an enhancement because it reduced the molten alloy's fluidity and made the casting process impossible. This was the reason to grade zinc as an impurity and to limit it to <0.01%.

Investigations that have been performed with cadmium concentrations below the solubility level have resulted in the white metal TEGO® V738 (5). Investigations at Th. Goldschmidt AG with cadmium and zinc above the solubility level failed. Therefore, further tests with zinc did not take place. However, in 1993 the author resumed the tests and found very similar conditions for cadmium and zinc. Below the solubility level, where no additional crystals form, a positive effect for the matrix can be found. Zinc is a full replacement for cadmium and has been more intensively investigated.

As we have here a multi-compound system, we manufactured, for an unambiguous assessment of the influences of the elements and concentrations on microstructure and compressive yield, first the alloys with the composition of the matrix phase, then the alloys with different contents of antimony and copper, and finally the alloys with the addition of different contents of zinc, measuring in each case the compressive yield $\sigma_{d0,2}$.

The values of the compressive yield of the alloys manufactured in the laboratory were all too high. This could be determined directly by comparison with the known values of alloys with similar compositions. The reason for this difference was the fact that very small quantities cast under laboratory conditions cool

Table 4.1 *Summary of Elements Soluble in Tin as Possible Replacements for Cadmium*

Element	Solubility in tin %	Remarks
As	20.0	toxic
Au	0.3	high priced
Bi	21.0	low melting eutectic with Sn
Ga	4.3	high reactivity, difficult to handle
Hg	1.0	toxic
Pb	2.5	low melting eutectic with Sn
Ti	0.6	toxic
In	10.0	toxic grading expected (pulmonary edema)
Zn	1.0	good properties, low cost

down under completely different conditions than plain bearing manufacturing conditions. The cooling in the laboratory is quicker, and the microstructure of the laboratory samples becomes considerably more compact-grained. That leads to better material properties. As the laboratory conditions remained constant, we could, in fact, determine through our examinations the changes caused by the individual element variations.

Zinc in the range of 0.3–1.0% results in almost the same increase of compressive yield rating as with 1.2% cadmium. Thus, zinc is a full replacement. The wide effective range is helpful for industrial use. In principle, zinc also has a refining effect but cannot be used in practice because of the limits on zinc concentration. It was therefore decided to add silver for refining. Now we see that the toxic elements cadmium, nickel, and arsenic have been replaced by zinc and silver. The new white metal is well known under the name TEGOSTAR™ (5).

A large-scale test program confirmed the feasibility both of manufacturing the alloy and casting it into plain bearings. After this success, the follow-up activities were concentrated on producing results for a technical data sheet. As the quality of an alloy is represented by the chemical composition and the solidified condition, it is important for the material tests that the specimens relate to practical conditions. That means specimens have to be selected with crystallization structure and size typical of bearing manufacture.

The standard tests for hardness, tensile, compressive, and rotating-bending strength were as expected, but the path of the stress-strain curve showed a noticeable linearity with a pronounced Hook area. This zinc-induced effect indicates a reduced plastic deformation of the new material.

The importance of creep resistance has been mentioned several times, but what is creep?

White metals usually tend to plastic deformation. In a plain bearing under constant high load and temperature, the white metal layer suffers a crack-free deformation in the direction of bearing motion. As a result, the geometry is changed, lubrication gap and lubrication supply are reduced, and damage occurs. Stable geometry is a precondition for long service life. To achieve long life of highly loaded plain bearings, a white metal with no tendency to creep must be used. The problem is that most of the soft white metals in general tend to creep. The high compressive capacity and reduced creep of TEGO® V738 (5) were realized by adding the element cadmium. For the new white metal, zinc has been used with even greater success.

Creep was not considered in the past, and there were no data available for the creep behavior of the different materials. To make possible some comparisons, we developed creep data as follows:

A white metal specimen of diameter 20 mm and length 20 mm is loaded at a constant 100°C with 15 N/mm² for 106 s. The creep at the end of the defined time period varies significantly for different white metals, as shown in Fig. 4.1. Compared with the standard alloy $SnSb_8Cu_4$ or ASTM B23-2, the zinc-hardened white metal creeps less by a factor of ten. Compared with the reference alloy, the factor is two. The creep of all other alloys falls between the values of the standard alloy $SnSb_8Cu_4$ (ASTM B23-2) and TEGO® V738 (5).

Thus, the zinc-hardened white metal has a better creep resistance than all other white metals by a significant margin. That is the fundamental reason for the success of this new material, and, therefore, it will be the basis for future white metal development. The higher a material is loaded, the more dimensional stability is needed.

The described tests refer to the solid white metal, and there are questions concerning whether the results would be similar for the material in compound with a backing shell and whether there is an adverse influence from the layer thickness. Therefore, additional long-term tests have been made on a compound specimen. The tests clearly showed certain trends: on a thinner layer the compressive yield strength increases and the tendency to creep falls dramatically. Thus, the results from Fig. 4.1 can be taken also for the material in compound.

Last, we carried out the impact-bending test. The test procedure is explained in the next chapter. Whereas on TEGO® V738 (5) it takes only 250 J of impact work to break the specimen, TEGOSTAR™ (5) takes a top position with 785 J. This zinc-hardened alloy can be allocated to the group of tough white metals and is able to withstand high dynamic stress. It is the first white metal to combine high static and dynamic capacities in one material and for universal use.

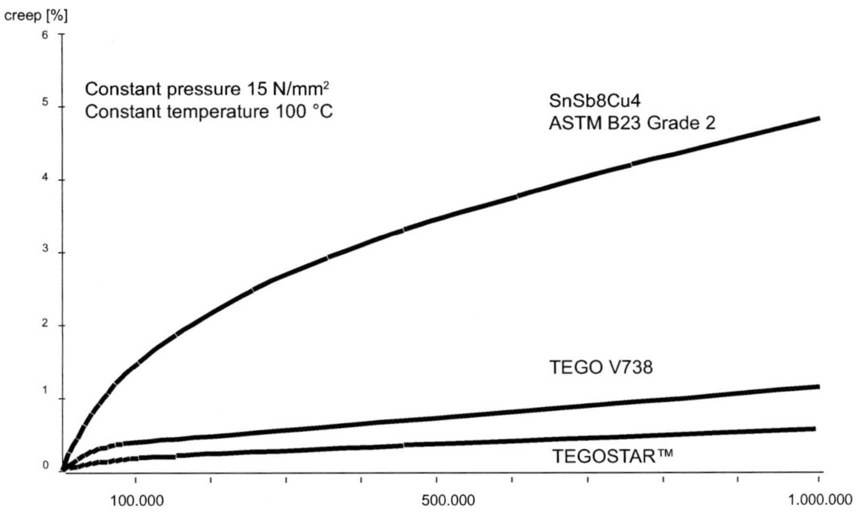

Fig. 4.1 Long-term creep in percent.

Of most interest is the behavior of the lining material at higher temperature. With a rising temperature, the drop in material properties is nearly linear, and because all white metals have a solidus of about 240°C, the values at room temperature indicate the capacity. The higher the compressive strength at room temperature, the higher the strength remaining at operating temperature. That means a significant advantage for the zinc-hardened white metal due to its extremely high compressive yield rating at room temperature. As shown in Fig. 4.2, TEGOSTAR™ (5) at an operating temperature of 100°C has a compressive yield strength of 50 N/mm², a higher figure than the standard alloy $SnSb_8Cu_4$ or ASTM B23-2 has at room temperature of 47 N/mm². That is a significant advantage for a plain bearing lined with TEGOSTAR™ (5).

The zinc-hardened white metal is not just an environmentally friendly alternative, but it also performs better than its predecessor. This is due to the combination, for the first time, of high-level static and dynamic capacity together with an enormous reduction in creep. The result is high loading capacity with long service life, well proven during the past ten years in many different industrial applications. This is the basic material preferred for further R & D with tin-based white metals.

4.5 Comparison of International Technical Data

White metals are in many countries referred to in national standards and specified with technical data sheets, which simplifies the making of comparisons. However, caution is advised here, because the American data basis differs from that of the rest of the world. When comparing the data of $SnSb_8Cu_4$ - DIN ISO 4381 with the data of ASTM B23-2 Grade 2, first it must be noted that there are different units in use.

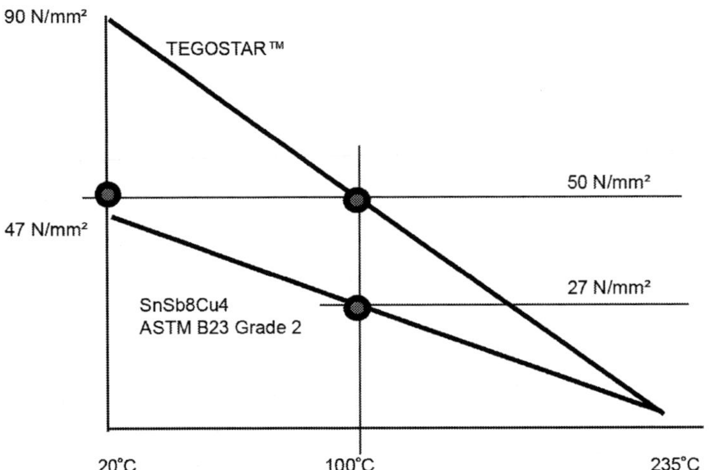

Fig. 4.2 *Compressive yield* $\sigma_{d0,2}$ *at various temperatures.*

DIN uses N/mm² and ASTM uses psi (1N/mm² = 145 psi). Even when converted, the data are not identical. This is due to different defined bases of measurement. In the United States, the strength data are based on 0.125% or 25% deformation; other countries refer to 0.2% or 50% deformation.

Because of the nonlinear curve of the stress-strain diagram, and the fact that the curves are not documented, the values cannot be converted and also cannot be compared. As a rough indication, it can be said that the American values for tensile and compressive strength are approximately 30% lower.

A similar problem occurs with regard to Brinell hardness. In Europe white metals are measured according to HBW 10/250/180, and in the United States they are measured according to 10/500/30. The ball size is the same, but the U.S. tests are with double load and 6× shorter time. The result is that in the United States, the apparent hardness has a different value, for tin-based white metals approximately 2 HBW more.

A direct comparison of U.S. material data sheets with those of other countries is not possible because of different test conditions.

4.6 Lining Materials of the Future

The material strength, especially the fatigue strength, cannot be derived simply from the tensile or compressive strength. It depends essentially on the creep behavior of the material. Therefore, it is reasonable and necessary to base future material developments upon the well-proven zinc-hardened white metal.

However, in the first instance an upgrade is not easy to achieve. On the one hand, the concentrations of the basic elements are limited by the applied lining processes. On the other hand, within these limits the possibilities have been largely exhausted.

It is clear that in the future, the development of improved plain bearing alloys is only possible when the restrictions from the lining process are eliminated. That means the first priority is to develop a new lining process that avoids the previous restrictions. Then the way will be free for unrestricted optimization of material, and the performance gap, between the white metals with their benign damage behavior and the bronzes with their better strength, can be closed.

The author has already started on this new development, and the results gained are very promising. They are presented in Chapter 9 in connection with the newly developed lining process.

The new lining process also allows lining with alloys in which homogeneous composites are embedded. That opens the potential for developing completely new lining materials for plain bearings.

Chapter 5

Compound Material, the Great Unknown

The quality of the lining material is not included in the design calculation of bearings. This fact means that in practice the two are not treated as having any connection.

By means of the design calculation, the operational safety should be optimized such that the bearing does not run hot or wear excessively. Design calculations assume that bearings have a limitless lifetime. If so in practice, then everybody is happy. If bearing damage occurs, it will usually be assumed that deviations of operational conditions are responsible.

Consequently, for many failures the symptoms are dealt with and the investigation of the underlying reason is neglected.

Fortunately, until now most plain bearings run without problems. But in the future, with rising requirements regarding bearing loadings, the situation may change. To meet these changes, it is imperative that in the future, attention be directed to the characteristics of the compound bearing material.

Plain bearing alloys have been tested very intensively in the past, but in most cases just the lining material alone, rarely in compound with the backing and hardly ever under real practical conditions.

5.1 The Material Data's Scope of Information

The duties that a bearing lined with white metal has to meet are focused on the loading in combination with the corresponding operational temperature and the resulting durability. For evaluation of the suitability of a plain bearing, usually the technical properties of the lining material will be used. In the past, it was never considered that the behavior of the lining material is completely different as a separate material than in compound with a backing. Therefore, the following important question has to be resolved: how much useful information is provided by the present material data? The following discussions will make the answer clear.

Up to now the documented technical properties of the lining material give no clear advice about performance of the plain bearing.

5.1.1 Technical Properties of the Lining Material

In bearing material standards and technical data sheets, the information is usually limited to chemical composition, tensile strength, and hardness.

Occasionally, additional data are given about compressive strength and dynamic capacity in the form of impact-bending, rotating-bending, or fatigue-bending strength.

5.1.1.1 Frictional Properties

On the technical data sheets, there is no information about frictional properties of the bearing alloys. Nevertheless, good frictional behavior will frequently be mentioned and promoted in advertising. Generally, it can be determined that all bearing alloys have good frictional behavior. For the lead- and tin-based alloys, this behavior will be mainly influenced by the matrix, and for the bronzes, by the lead content. The variations inside an alloy group are very little. Furthermore, the frictional properties are not of primary concern for bearings, as they operate hydrodynamically, and during the mixed lubrication situation during starting and stopping of the machine, none of the bearing alloys gives any problems. Bearing damage due, for example, to lack of oil proceeds very quickly, and whatever the frictional properties, they cannot save the situation. Hence, the frictional properties are no criteria for bearing material selection.

There is no need for further investigation of frictional properties of bearing alloys, as they are well proven for mixed friction conditions, and they are well known for benign failure behavior without damaging the shaft.

All well-proven bearing alloys have good frictional behavior.

5.1.1.2 Tensile Strength

Tensile-strength figures are not reliable indicators of bearing quality. The inner layer of a plain bearing will be subjected to compression load, and the compression capacity for most bearing materials is higher than the tension capacity.

The ultimate tensile strength of the lining indicates solely the maximum achievable bond strength and, in conjunction with a destructive bond test, aids control of the bearing production process.

The ultimate tensile strength of a bearing alloy is not of interest apart from evaluation of bond tests.

5.1.1.3 Compressive Yield Strength

The value of the 0.2% compressive yield strength depends on temperature and layer thickness. At higher temperature levels, the value is further reduced by lead impurities. For accurate comparison of different materials in general, the data should be based on testing at 100°C. Data sheet figures generally refer to the solid material. The influence of the compound of layer and backing materials, and especially the influence of the layer thickness, are not covered.

The compressive yield strength, as documented in the material data sheets, does not include influences of compound and layer thickness.

5.1.1.4 Compressive Yield Strength in Compound

In the past, there have been investigations [3] to discover whether, and to what degree, the lining of a plain bearing benefits by the compound with (i.e., being bonded to) the backing. It has been found that the compressive yield strength $\sigma_{d\,0,2}$ increases as the lining thickness decreases. That is clearly plausible because the thinner the lining layer, the more the influence of support from the backing.

Unfortunately, the recorded curves from the original tests were spread over such a wide range that they were unusable in practice. From practical experience, for example with bearings of steel mills, it is well known that the higher compressive yield strength due to decreased layer thickness can be quite useful.

The author studied this phenomenon further. The first task was to establish why the earlier test results spread over such a wide range. It became obvious that the results had been influenced by friction effects during the test. The compression stress generated a lateral strain on the white metal layer of the specimen. This strain was more or less restricted by contact with the compression cylinder. With the new tests, the interference from strain friction was eliminated by using two compound specimens with the white metal layers face to face. Both specimens received the same load and the layers the same strain. So there was then no obstruction influencing the results of measurement.

The new test method gave clear curves, as shown in Fig. 5.1. The value of compressive yield, given on the data sheets for the solid materials, will be found in the curve for layers thicker than 5 mm. That means for thicknesses below 5 mm, in compound, the compressive yield increases considerably.

For a layer thickness of 1.5 mm, the value is nearly doubled and for 0.5 mm, more than tripled. Further investigation has established the mathematical relationship [4].

The described investigation delivers, for the first time, data considering the compound. For practical use, the influence of the increased compressive capacity on the durability of a bearing must be clarified. And, of course, to use a higher compressive load, the bearing design has to be calculated for the new parameters, because there are modified influences on the hydrodynamic conditions.

With determination of compressive yield in compound, the associated durability of the plain bearing remains to be established.

For increasing load, the compound material and the hydrodynamic conditions have to be examined and adapted.

5.1.1.5 Hardness

The hardness test on solid white metals will be based on DIN ISO 4384-2. For all tin-based white metals, a high tin content, in the range of 80–89%, gives a

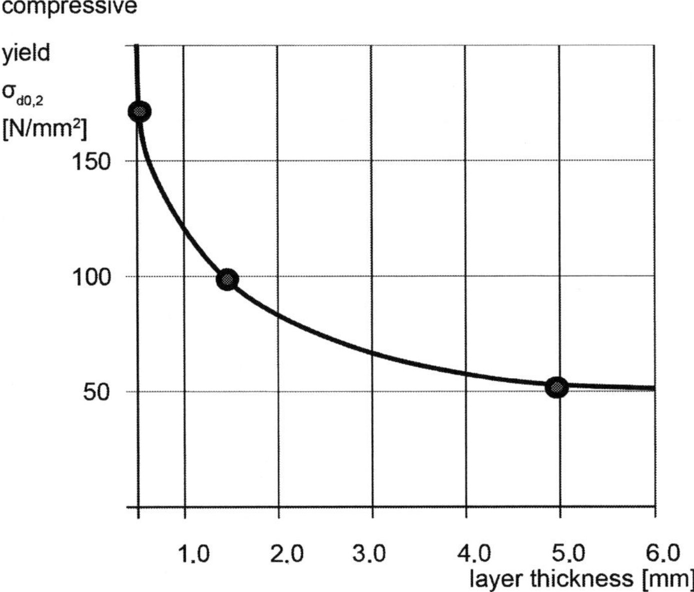

Fig. 5.1 *Compressive yield $\sigma_{d0,2}$ of the white metal TEGOSTAR™ (5) at 100°C, corresponding to the layer thickness in compound.*

hardness result on a uniform level of 23–35 HBW at 20°C and 10–17 HBW at 100°C. The low hardness allows good embedding of foreign material in the lining surface and, in particular, enables a failure behavior that is progressive and avoids damaging the rotor. The hardness values of the different tin-based white metals, when compared, do not lead to a quality conclusion and are, therefore, not differentiating factors.

Newly developed white metals with a higher copper content have, at room temperature, considerably higher hardness data. But based on the tin matrix, there is no significant change of solidus, and therefore the hardness at higher temperatures declines to the level of the other white metals. That means there is no significant hardness difference under operational conditions between the white metals containing high and low copper.

The hardness test on material compounds regularly returns higher results than those documented for the solid material. That is a result of the influence of the backing and the layer thickness. We have seen already that hardness tests on solid material provide no quality conclusions. Likewise, the hardness tests on the compound materials give no useful information, as no reference data for the compound in relation to the layer thickness are available.

That is the reason for the author's initiative of canceling the hardness tests on compound bearings with tin- and lead-based white metals under ISO 4384 part 1.

Hardness tests do not provide a quality conclusion for the layer material or the plain bearing.

5.1.1.6 Impact-Bending Strength

The impact-bending strength is a numerical value to show different levels of impact resistance. It gives a rough indication for rating the different materials. It is a test procedure developed at Th. Goldschmidt AG (see Fig. 5.2). During the test, a hammer falls on a 10-mm white metal rod and delivers a radial impact at the midpoint between supports. After each impact, the rod is rotated 72°. The hammer creates impact energy of 0.275 J. The test measures the number of impacts until the rod breaks. From that the impact energy can be deduced.

The test is applied to the solid white metal, and therefore these results also do not consider influences from a compound with a backing material. Furthermore, plain bearings are rarely subjected to intensive impacts. So the test merely gives indications for comparison of the toughness of different materials. Unfortunately, data sheets list only data at room temperature. At higher temperatures, the impact-bending test results generally decrease, but there is a more significant decrease for alloys with lead impurities.

The impact-bending test does not consider influences from compound construction. The test simply gives rough indications on toughness of different lining materials.

5.1.1.7 Rotating-Bending Strength

The test is based on DIN 50113, and this again is performed on a solid rod of white metal. The bending creates tensile and compressive stress along

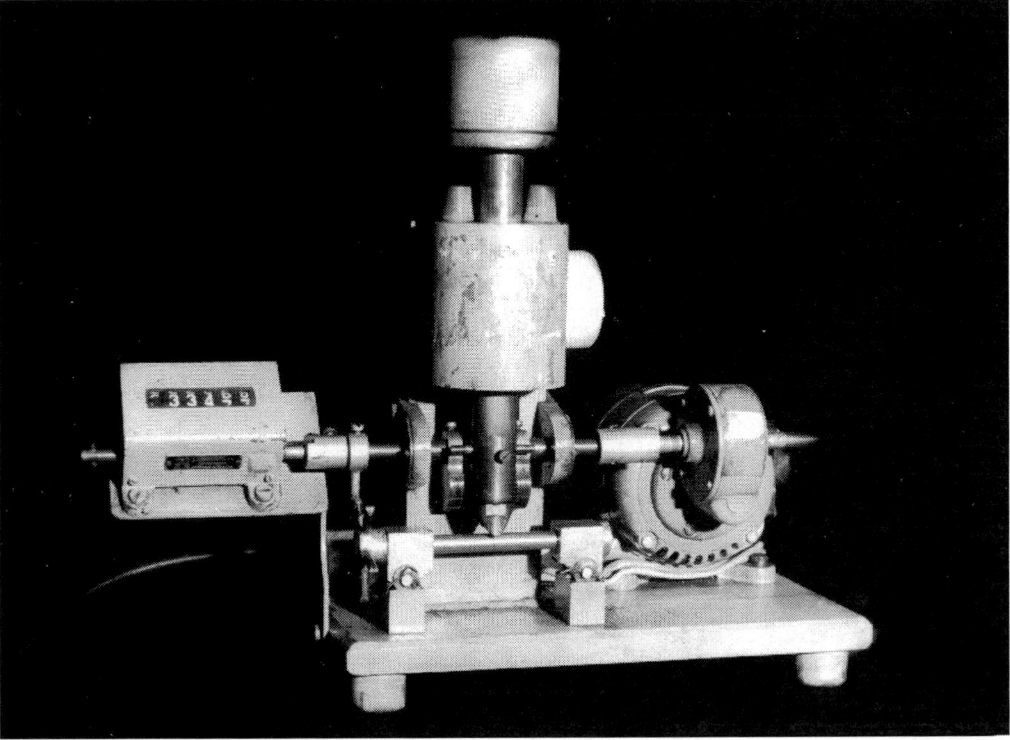

***Fig. 5.2** Equipment for testing the impact-bending strength.*

the cylindrical surface of the rod. The rotating-bending strength is based on the maximum tensile force the material can endure without failure at room temperature for 10^7 load cycles.

This is another test on solid lining material without consideration of influences from a compound structure. Even when similar tests are occasionally done on flat compound test pieces to DIN ISO 7905-3, the nature of applied stress, its direction, and the operating temperature remain unconsidered.

The standardized rotating-bending test is not applicable for plain bearings because their specific boundary conditions are not considered.

5.1.1.8 Dynamic Strength

The dynamic strength test based on DIN ISO 7905-2 is made on a solid rod. It applies an axial tensile and compressive pulsating stress for 10^7 load cycles. Using this test, Smith Diagrams [5] have been documented for many white metals per DIN 50100.

From the Smith diagram (see Figs. 5.3 and 5.4), data can be taken for pulsating tensile stress, pulsating compressive stress, and alternating stress, for which the alternating stress is identical with the rotating bending stress. Smith diagrams have been documented for white metals at 20°C and 100°C. But again these data cannot be used for a plain bearing compound because it uses a test piece of solid bearing metal.

The standard dynamic strength test is not applicable to plain bearings because their bi-metal construction is not considered.

5.2 Failure Criteria of Material Testing

In connection with the many dynamic strength tests on white metals, there were also extensive investigations into the fatigue behavior of white metals [6, 7, 8]. On all the tensile, compressive, bending, and pulsating tests that we have discussed, tensile and compressive stresses along the specimen surface, and inside the specimen parallel to the surface, will be produced. Surface dislocations occur in the direction of these stresses, and fine-flaked, linear-structured, or wave-structured deformations develop at the surface. Later micro cracks start, beginning on the fatigue glide bands, borders of crystals, or inclusions. Eventually the material fails.

The fatigue damage always starts on the specimen's surface that is parallel to the applied stress. No starting cracks have been detected inside the specimens [8].

5.3 Failure Criteria of the Plain Bearing

For many decades, the same standardized tests have been performed on white metals as are generally in use for all other materials. During all these tests,

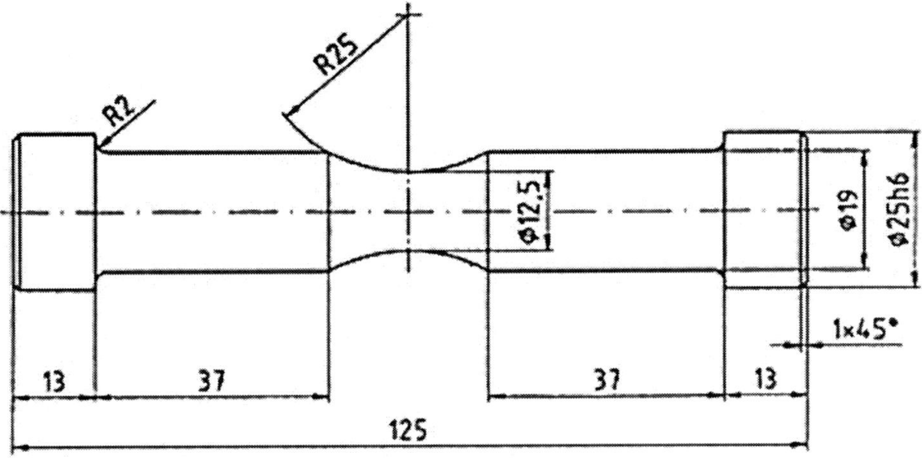

Fig. 5.3 *Solid specimen for the dynamic strength test.*

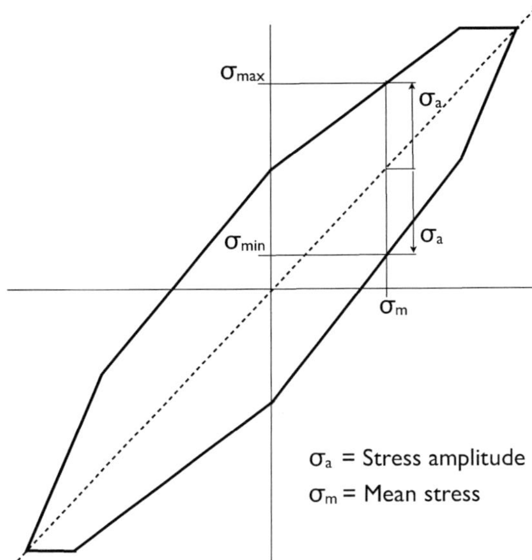

Fig. 5.4 *Smith diagram.*

nobody considered that for the loaded area of a compound plain bearing conditions are different: There is no surface parallel to the main stress!

Consequently, the mechanism of stress and results as found on a standardized specimen under bending and pulsating operation are not transferable to the compound-construction plain bearing.

All previous material tests on plain bearing alloys are based on classic methods. It has not been considered that the compound materials of a plain bearing are subjected to completely different conditions. For that reason, all the existing data are unsuitable for the assessment of plain bearing capacities.

The priority challenge is the investigation of the characteristics of compound materials. They control the permissible load level and the service life of the plain bearing. It is urgent and essential to identify the interactions of load, temperature, and stiffness. New test procedures for compound materials have to be developed. The compound material must no longer be the big unknown. This new awareness will affect the design and calculation of both present compound materials with high capacity, and new future developments.

There is need for further research on compound materials.

Chapter 6

Investigations on Test Rigs

Time after time, plain bearing problems are investigated on test rigs and the results compared to design calculations and to practical experience. Evaluation of the test results will include a confident assessment of the adequacy of the white metal despite the fact that, until now, no data have been available about the properties of the white metal when lined onto a backing shell to form a compound bearing material.

Previously, this observation would have been dismissed as misguided and inapplicable. Hopefully, the earlier chapters of this book will have given the reader reason to question the old assumptions and raised some thought-provoking contradictions. As the book progresses, we will discuss in more detail the compound bearing materials and compare properties with existing material data lists. This exercise will clearly show the lack of conformity and absence of knowledge about bearing materials when bonded to a backing layer to form a compound structure.

6.1 A Retrospective on the History of Test Rig Design

Plain bearings have been tested countless times on many designs of test rig. A historical retrospective proves to be interesting and instructive.

Gröber [9] wrote in 1921, "It is not sufficient just to do tests on test rigs. Such tests without a good theoretical basis merely result in an increase of already bulky literature but without academic progress and without practical numerical values." Vogelpohl [1] in 1967 re-quoted this statement of Gröber, obviously with some reasons.

Schmid and Weber [10] in 1953 stated the following regarding tests: "An authentic assessment of a plain bearing material for a special application will only be given when the bearing is fitted in its operational location. Although this procedure can be realized in single case, it cannot be the rule. This procedure is also not applicable for developing plain bearing materials further, because it does not allow a global assessment of the materials. Test procedures have to be developed that allow, with short time tests, numerous results for the tested bearings. The most important precondition for these tests is developing numerous ascertainable properties that describe the running conditions for all practical applications. For testing white metals at present it is in no way fulfilled."

In the same book, Schmid and Weber document 52 existing rigs for static, dynamic, and wear tests.

Vogelpohl [1] wrote in 1967, "The absence of basic rules, and the persistent habit of the white metal producers to impress with artificially high data, results today in demands which in practice are out of the question. Many tests with plain bearings have been made in over simplified conditions. Consequently the results contribute nothing to our knowledge."

Since 1998, DIN ISO 7905 Part 1 *Plain bearings in test rigs and in applications under conditions of hydrodynamic lubrication,* has been available. Remark 1 observes: "The diversity of practical applications with their different requirements leads to the construction of many test rigs. If the conditions modelling the lubrication are not defined in all details, the results from different test rigs generally are neither comparable nor transferable to practical applications. Different test rigs can lead to contradictory assessments of the same plain bearing metals."

Since 1953, the number of test facilities has greatly increased, but comparison of reports from the 1920s to 1950s with the current standards shows very emphatically that the old warnings and caveats have been ignored. There has, therefore, been no appreciable progress, and the problems remain unsolved.

A good example of this lack of progress is ISO 7905 part 2: here tests with massive cylindrical specimens are discussed. It is explicitly suggested that the procedure is for testing the fatigue of white metals that are not in compound with a backing metal. But this is exactly the problem, because basic experience with compound material is missing. The properties of bearing materials lined onto a backing differ from those of the solid white metals, which are given in the data sheets, and so the existing information is no help in accurate assessment of fatigue strength of a plain bearing, as already shown in the previous chapter.

As long as the characteristics of materials in compound are unknown and not considered, it is pointless to run more plain bearings on test machines. The results cannot be authentic or generally valid.

Anyone involved in developing improved bearings should consider the mentioned quotations of Gröber, Vogelpohl, Schmid, and Weber, even though they date from around 90 years ago.

The largely unknown influence of the backing shell is one of the reasons why test bench results for compound bearings differ widely. A further reason is the over simplification of the test conditions to the extent that the real application is no longer represented.

Within the last few years, the author has come to the conclusion that without broad knowledge about the compound of bearing and backing materials, further improvement of plain bearings will be unlikely.

Fatigue behavior data of the material when in a compound format provides the basis for a safe design.

Future investigations of bearing material in compound with backing layers must consider the true operating conditions when selecting the test procedure and loading parameters, and during the test procedure itself. So far, this has not been done consistently. Certainly, it is a subject of great complexity, and to put into practice all important conditions will create problems in the future. But to identify and to address the problem is already a step in the right direction. We must no longer be content to see the test procedures of the last eighty years as absolute valid standards or to repeat them with all their shortcomings.

6.2 The New Procedure for Testing Compound Bearings

The realization of the inadequacy of the existing data on the properties of materials for lining plain bearings, and the trend toward higher loading, led the author to develop a new test procedure that is closer to the real conditions of plain bearings. It is the first method able to produce data on the compound material, which is long overdue as a basis for designing highly loaded plain bearings.

The test arrangement is explained in detail here. Hopefully it will present ideas for future similar activities.

The issue to be solved is as follows: a compound bearing material consisting of backing shell and lining layer has to be tested for fatigue resistance under conditions representing actual plain bearing operation.

The relationship between load and temperature is a fundamental consideration. Better material leads, in principle, to better fatigue results, but with higher loading, the temperature rises and the fatigue capacity is reduced.

The new test arrangement is based on the following conditions:

- The applied load is a pulsating compressive load acting perpendicular (at 90°) to the running surface.
- The load transfer is not by direct contact but via the hydrodynamically generated lubricating film.
- There is no wear, and the associated temperature is generated automatically.
- The sliding speed is a variable influencing temperature and fatigue.
- The stiffness of the compound structure has an influence on fatigue.
- The arrangement is simple but as close as possible to practical conditions.
- Test conditions are designed to produce results in a relatively short time.

The concept for the fatigue test rig developed by the author is based on the principle of a thrust-bearing system. Usually in the set-up for a thrust bearing, the thrust collar is made of steel and the pads are lined with bearing metal. This is correct for practical use because the pressure parabola developing on the surface of each pad constitutes a static stress for the pad and its lining. As a reaction, a pressure parabola of the same shape and size arises on the thrust collar. For the rotating thrust collar, the pressure parabola consists of a dynamically circulating stress. For a journal bearing, the conditions are comparable so that there, too, the bore is lined and not the shaft.

A journal bearing develops just one pressure parabola, whereas a thrust collar has multiple pressure parabolas, one for each pad. Here there is a higher frequency of load cycles because the pulsating stress around the thrust collar is multiplied by the number of pads. This effect provides quicker results on the test rig, where the lined bearing test piece replaces the thrust collar and the pads are simply pure steel (see Fig. 6.1). Circular pads are used, as this is the ideal shape to avoid unwanted influences.

The test assembly is located at the top end of a vertical pinion shaft, and there the test disc is clamped in position, as shown in Fig. 6.2. The test disc can be changed quickly from test to test without draining the oil. The test rig has three circular steel pads that are uniformly loaded against the test disc by a hydraulic system. (Figure 6.3 illustrates a cross section of the test rig, while Fig. 6.4 shows assembly of a test disc.) The frequency of rotation is set at 48 Hz, which with the three pressure pads gives a load cycle frequency on the test material of 144 Hz. At this frequency, ten million load cycles will be completed after approximately twenty hours. The standard settings provide for a peripheral speed of 12 m/s, oil of 220 ISO VG, and inlet temperature of 50°C. Figure 6.5 shows a test disc after the procedure; Fig. 6.6 presents typical fatigue damage. Figure 6.7 shows a Koring fatigue test rig.

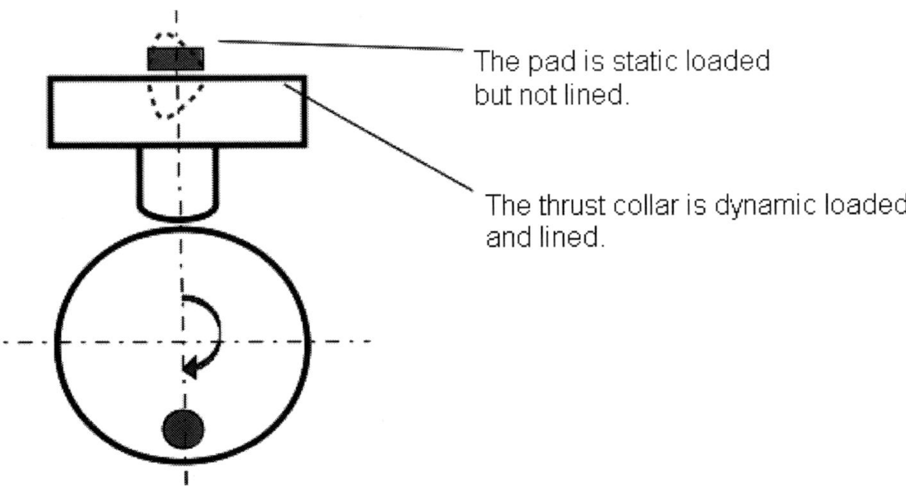

Fig. 6.1 *Principle of the Koring fatigue test rig.*

6.2.1 Summary

Very soon it was clear that for determining fatigue capacity, the maximum pressure p_{max} is the controlling value and not the average pressure. That is plausible because p_{max} is many times greater and the associated temperature is also at a maximum. The damage always begins in the area of maximum load in combination with maximum temperature.

Interestingly, p_{max} is not constantly proportional to the average pressure. The relationship depends on the load level and on the geometry of the bearing. With an asymmetric load distribution, p_{max} increases further, and on circular pads it may be nearly 4× the average pressure.

For fatigue strength of plain bearings, p_{max} is the reference and not the average pressure.

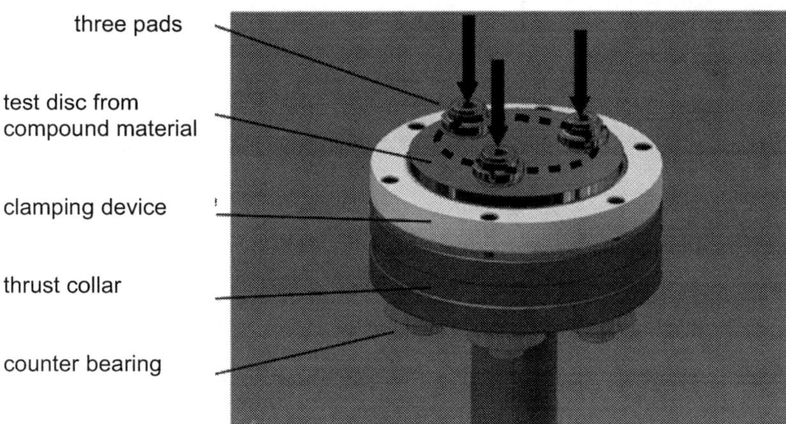

Fig. 6.2 *The new rig for fatigue tests on plain bearing compound materials.*

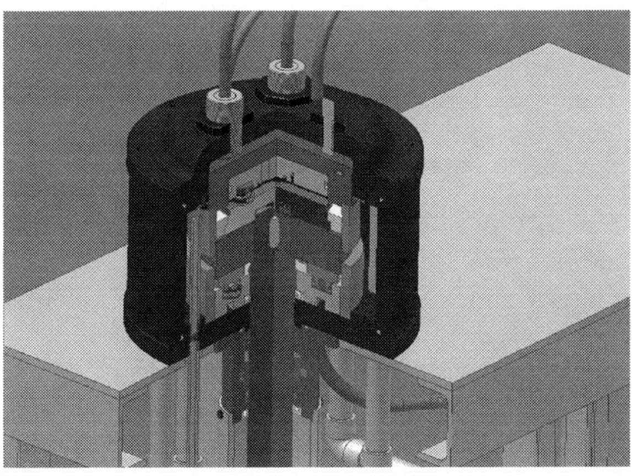

Fig. 6.3 *Cross section of the test rig.*

Fig. 6.4 *Assembling a new test disc.*

Fig. 6.5 *Test disc after the test procedure.*

There remains the critical question of the extent to which the maximum temperature in the lubrication gap is identical with the maximum temperature at the material surface. For a low-load-cycle frequency, the material temperature will certainly be somewhat below the temperature of the lubricant, because each load cycle will vary the lubrication gap and the bearing briefly gets more cooling oil. At higher frequencies, as on the test rig, the two temperatures are very close. The temperature in the lubrication gap can be measured and calculated with high accuracy, and so we can take it as the reference temperature. For any bearings with a low-load-cycle frequency, it will be on the safe side.

Using the new test rig, fatigue data can be established within a relatively short time. The relationship between fatigue and temperature will be considered automatically, and in the future the fatigue strength will be documented with the associated temperature. Since the temperature is also influenced by the sliding

Fig. 6.6 *Typical fatigue damage.*

Fig. 6.7 *General view of the Koring fatigue test rig.*

speed, the fatigue-temperature values have to be documented for different sliding speeds.

The new test rig permits checks such as whether the fatigue strength depends on the material layer thickness. During the tests, it has been found that the stiffness of the compound material has significant influence on the fatigue performance.

Valid data about fatigue capacity of the compound will be needed as a basis for designing durable plain bearings.

With the new test procedure presented here, the way forward is described and must be followed to gain the benefit of accurate data in the future.

Chapter 7

New Conclusions Relating to Compound Materials for Plain Bearings (6)

7.1 Performance Ranking of the Compound Materials

In Chapter 5 it has been established that the material properties available up to now have no informative value for establishing fatigue strength of a plain bearing. Comparison of these data with the test results from the new fatigue test procedure on compound materials confirms that clearly.

7.1.1 Ranking by the Fatigue Strength of the Compound Materials

Table 7.1 shows a selection of well-proven white metals with their technical properties. In first place are the previously described all-around alloys TEGOSTAR™ (5) and the laser-lining version LACOMET® (5). In second place are the alloys HM07 and HOYT® 11R (5), which are well proven in the marine industry, followed by the standard alloys $SnSb_8Cu_4$ based on DIN 4381, identical with ASTM B23 Grade 2. In fourth place is the cadmium-improved alloy TEGO® V738 (5), and in fifth place the well-known alloy LM THERMIT, representing all lead-based alloys. For comparison purposes, a further variation of LACOMET® (5) with higher copper content is listed. This shows clearly the development potential of laser-lining. The ranking order follows the results from the new fatigue test procedure and confirms the experience from practice.

***Table 7.1** White Metals Ranked by Fatigue Strength and Listing Other Technical Properties*

		Fatigue limit p_{max} and associated temperature at 12 m/s peripheral speed by Koring [N/mm², °C]	0.2% compressive yield by DIN [N/mm² at 100°C]	Impact-bending strength by Goldschmidt [Average impacts to failure at 20°C]	Rotating-bending strength by DIN [N/mm² at 20°C]
	LACOMET® with extended Cu content	142 N/mm², 112°C	77	**	**
1	TEGOSTAR™ - LACOMET®	104 - 113 N/mm², 106°C	50 - 58	2856	±35
2	HM07, HOYT® 11R	99 N/mm², 105°C	43	2689	±33
3	TEGOTENAX, ASTM B23 Grade 2	86 N/mm², 103°C	27	3741	±29
4	TEGO® V738	32 N/mm², 93°C	48	910	±39
5	Thermit (Lead-based alloy)	**	27	285	±28

** not tested

7.1.2 Rank in Order of Compressive Yield Strength

Based on the compressive yield strength data, the rank order in Table 7.1 would be 1-4-2-3-5. Obviously, this does not match the fatigue strength order. From TEGO® V738 (5), we have practical experience that the material has poor capacity for withstanding dynamic load. It would be seriously overrated here. Therefore, compressive strength alone gives no hint of the fatigue capacity of a compound material.

7.1.3 Rank in Order of Impact-Bending Strength

Listing the fatigue capacity based on impact-bending data, the rank order would be different again, namely 3-1-2-4-5. On this basis, $SnSb_8Cu_4$ would take first place. This material has unquestioned good resistance to dynamic loads. But we know from practical experience that the material is below the capacity of TEGOSTAR™ (5) and HM07. The obvious reason is the poor creep behavior of $SnSb_8Cu_4$ together with its low compressive strength.

7.1.4 Rank in Order of Rotating-Bending Strength

The rotating-bending strength results of all white metals are close together in a range of ±28 to ±39 N/mm². Sorted by these values, the rank order in Table 7.1 would be 4–1–2–3–5. Here the value of the tin alloy $SnSb_8Cu_4$ is nearly the same as the value for the lead-alloy. This circumstance has already been noted and given low credence, because practical experience is that lead-alloys have a significantly lower overall quality level than tin alloys. TEGO® V738 (5) is in first place of the rotating strength rank order, but 60 years' experience with this material has demonstrated that it has inferior resistance to alternating stress compared with other materials. It is clearly behind $SnSb_8Cu_4$. Obviously, the rotating-bending strength data give no indication of fatigue capacity of a plain bearing.

7.1.5 Summary of Rank Order Based on Different Properties

Fatigue strength of the material	1-2-3-4-5
Compressive strength of the material	1-4-2-3-5
Impact-bending strength of the material	3-1-2-4-5
Rotating-bending strength of the material	4-1-2-3-5

This comparison of rank orders based on different properties shows dramatically that the material properties available until now lead to extremely different and contradictory results: a situation that results from the inadequate conditions of material tests, as described in Chapter 5.

Properties of a solid material are not identical with the properties of the same material in a compound structure.

Only the new fatigue test—based on conditions emulating the practical use—leads to results with high conformity to practical experience.

7.2 Fatigue Strength According to the Layer Thickness

Chapter 5 described the way the compressive strength increases when the layer thickness decreases. Furthermore, the tendency to creep declines with reduced layer thickness. At first, this finding could lead to the speculation that fatigue behavior follows similar rules.

Compound specimens with different layer thicknesses ranging from 2 to 0.5 mm have been tested, and no variation of fatigue capacity has been found. This confirms earlier intensive investigation on automotive plain bearings with very thin layers [11]. The result there was that a detectable fatigue capacity increase only occurs on layers thinner than 0.25 mm.

Only extremely thin compound layers bring an increase of fatigue capacity.

7.3 Observations on Thin Layers of Tin Alloys

During the tests with a 0.5-mm layer in compound, the following has been observed: after long exposure to a pulsing high load, the surface becomes modified. From a smooth plane, it changes to texture similar to sharkskin or a golf ball structure. Clearly there is a compression of the lining layer, which is affecting the matrix more than the embedded crystals, hence the surface texture. This compression differential increases as the layer thickness decreases.

A drop in the oil film operating temperature of approximately 5% has been observed. On the one hand, it is desirable to use this effect for reducing temperature and friction. On the other hand, the texture is so little that it will be lost with a mixed-friction condition during start-up or stopping of the bearing. Thus, a permanent advantage cannot be assumed.

7.4 Influences of Peripheral Speed and Frequency of Load Cycles

At first, all tests were made at a peripheral speed of 12 m/s and a frequency of load cycles of approximately 150 Hz. The tests were then repeated with 4 m/s and approximately 50 Hz. Following this modification, the temperature dropped by up to 35°C with the result that the fatigue strength improved by approximately 13%.

With rising peripheral speed and temperature, the fatigue capacity drops.

7.5 The Limits of the Measurable Fatigue Strength

It appears that an increase of applied load cannot be made pro rata with the factor of material improvement. In general, with each decision for applying higher load, the bearing first has to be recalculated for the new conditions to be sure that the hydrodynamic properties are fulfilled. The range of fatigue resistance of the compound material is limited, and beyond the limits there are other damage characteristics. The lower limit is given by mixed friction based on high load, low speed, and low temperature. The upper limit is also characterized by mixed friction. Here high load and high speed bring an increased temperature with diminishing oil film gap. For premium compound materials, the bearing may enter into mixed friction before the material fatigue resistance can be measured.

The measurable and serviceable fatigue resistance of a bearing is limited by mixed-friction factors.

7.6 Fatigue Resistance of the Compound Material and the Influence of Stiffness of the Plain Bearing

In general, the fatigue test results are at a very high level. They are much higher than the usual plain bearings' load level. Still, in practice, plain bearings can suffer damage from dynamic overload. Why is that?

The test results are based on conditions under which the test specimen is supported with infinite stiffness. The test specimen is subject only to the pulsing compressive load with no additional stresses from any deformation.

On a plain bearing, in practice the conditions are different. On each load cycle there is an additional stress coming from bending deformation of the bearing. Consequently, the fatigue resistance will be dramatically reduced. The amount of additional bending stress depends on the stiffness of bearing and housing.

Reduction of the stiffness of the journal bearing and the housing results in reduction of the fatigue strength of the bearing.

Further tests to determine the interaction between plain bearing stiffness and compound material fatigue resistance are made with the test specimens no longer fully supported but set on three joints, as shown in Fig. 7.1. So the circumferential load applies as if on an endless continuous beam. Supporting conditions and field width of the beam has been optimized by FEM calculation so that an additional bending stress of 11% of the operating compressive stress is applied.

This proportion equates to the typical stiffness and stress for bearings in industrial use and takes into account both bearing and housing stiffness. It is planned to repeat the tests with other stiffness parameters. Therefore, we have data becoming available to determine fatigue resistance of a bearing considering its stiffness and the applied load.

7.7 The Deformations of a Plain Bearing

We are at the very beginning of these investigations. It should, in the future, be helpful to assess the various deformation influences by considering each of them

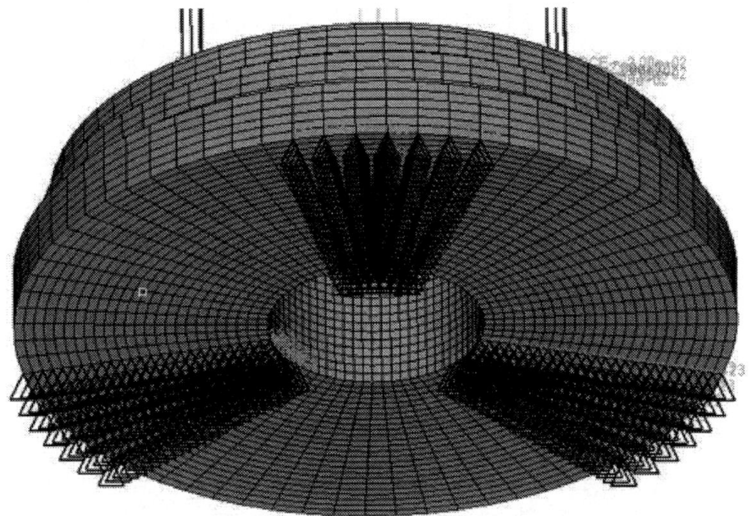

Fig. 7.1 *Location of supports underneath the test specimen as three triangle segments. Bending stress equates to loads on an endless continuous beam. In practice, this bending stress is overlaid on the applying compressive load.*

separately. They will be generated in different ways and they appear differently. There are elastic deformations and permanent deformations.

7.7.1 Elastohydrodynamic Deformation (EHD) of Plain Bearings

Two different approaches are in use for calculating plain bearings: the older hydrodynamic approach (HD), which refers to a rigid geometry, and the elastohydrodynamic approach (EHD), which is used increasingly more. Besides the nonlinear stiffness and damping coefficient of the oil film, the EHD method considers the elastic deformation of the bearing and the resulting conformation between bearing and shaft. This is especially useful for calculating connecting rod bearings for which the aim is to reduce the maximum stress peak in the bearing and thus to lower the mixed friction risk. Deformation of both bearing and shaft due to load and temperature are considered and, by modifying the stiffness, misalignment will be reduced and the oil gap geometry optimized. With good conformity between shaft and bearing, the pressure distribution is improved and the maximum pressure is reduced with the aim of achieving high operating safety and longer availability.

This has been the situation since around 1974 [12, 13], and, although today's computers are much more powerful and calculations are more detailed, the basic parameters of design have remained unchanged for all these years [14] and fail to account for an important fact. The influence of having a compound material is not included in the calculation parameters, and in fact the compound material is not considered at all.

For static loading, the existing EHD method may give an optimal result. For dynamic loading, it is completely different. The more elastic the bearing design, the more additional bending stress applies to the compound material. The fatigue resistance becomes drastically reduced. Because of the negative effect on the compound material, for further EHD optimization the compound material and its fatigue behavior must necessarily be considered. Otherwise the problem of untimely failure will intensify.

It is essential to include the fatigue behavior of the compound material into the EHD approach.

The elastic deformation of a plain bearing is given by the stiffness of the bearing shell and the stiffness of the housing in which the bearing is assembled. The proper fitting between these components is very important. If the fit is not correct then the bearing shell suffers greater deformation and increased bending stress. Practical experience is that this leads within a very short time to a damaged bearing.

This practical experience gives a clear hint that with increasing bending stress the compound bearing loses fatigue resistance.

With the EHD effect, it was assumed, up to now, that the service life can be increased solely by consideration and adoption of stiffness with its corresponding reduction of the maximum pressure. The compound material and its dependency on the component stiffness have not been considered. Therefore, until now only a limited improvement of the total condition was possible. For future optimization, the interaction between deformation and fatigue resistance of the compound material has to be considered.

The example shows clearly how complicated the connection is. Some investigators are busy with application technology problems of plain bearings and optimize the deformation [14]. Others are working to determine the deformation stresses and find their influence on early failure [15]. Material specialists have very often analyzed the lining material based on many methods [6, 7, 8]. Lack of cooperation between the different disciplines and their typical interface problems have resulted in a lack of documented useful results.

Intensive interdisciplinary cooperation of research and development activities on plain bearings is the precondition for success.

If, on a research work, an important basic condition is not registered, then its influence is not taken into account. The consequence is an inaccurate result.

A reduction of bearing stiffness results in a reduction of fatigue strength of the compound layer.

7.7.2 Different Deformation Types of the Lining

Deformation affects not only the housing and bearing shell but also the lining material. When considering the effect of deformation on the function of a bearing, it becomes necessary to assess the different types of deformation. In the future, a better understanding of the different deformation types will lead to more accurate calculation results and improved compliance with test results.

7.7.3 Elastic Deformation

A compressively loaded surface experiences an elastic deformation of the layer in proportion to the compressive load. This effect also leads to a local conformation between bearing and shaft.

7.7.4 Deformation Due to Material Compaction

The layer in the area of maximum compression load undergoes compaction. In contrast to the creep effect, the compaction deformation is static. The amount of compaction depends on the loading level, the type of lining material, and the thickness of the lining.

7.7.5 Modification of Surface by Mixed Friction

After multiple starts with short periods of mixed friction, there is at first a smoothing of the surface in the area of maximal load. The roughness peaks will be removed.

Deformation from mixed friction and deformation from compaction together compose the polished contact area. It becomes visible on the bearing surface after some service time. For a radial bearing, this deformation leads to a better conformity of bearing to shaft.

7.7.6 Creep Deformation

The very soft tin-based lining materials trend to creep, especially when the tin content is above 82% and no elements are added to the alloy specifically for creep reduction; see Chapter 4. Creeping consists of a crack-free plastic deformation of the layer in the running direction of the bearing under high compressive load and high temperature. Consequently, there is a change of the bearing geometry. The result is a bearing damage. As long as the load and temperature apply, the creep progresses. Materials that tend to creep should not be selected for high load bearings.

Recognition that different types of deformation act on a plain bearing will bring better results.

Layer materials that tend to creep are not fatigue safe.

7.7.7 Deformation Due to Lining

Deformation resulting from lining becomes clearly visible when splitting bushings into two halves. After splitting, an increased curvature will be found. Independent of the size of the bearing, the diameter will be reduced by up to 3 mm, measured from joint face to joint face.

This is commonly explained as being the bimetallic effect: the strain modulus of the white metal is double compared with the strain modulus of the steel backing. Therefore, during cooling, the lining shrinks twice as much as the steel and causes the deformation of the half shell.

The explanation is plausible. But is it automatically also correct? Obviously, in the past there were no doubts raised, and therefore the phenomenon has never been tested.

A plausible explanation is not always the correct one.

Just now we started to investigate the circumstances in detail. The intermediate results already clearly show that the deforming force does not come from the lining but from the backing. The reasons will be determined now and the author will report about that later. At least the discovery is comforting because, if the previous assumption was correct, every bearing would have high-tension stress in the lining after casting. When the backing was deformed by truing activities back to the original shape, then the lining would be under double tensile stress. That would be a big uncertainty factor for the service life of the lining. Practical experience teaches us that this obviously is not the case.

7.8 Defects in the Layer

The fatigue tests with compound specimens have been made with 2-mm, 1-mm, and 0.5-mm bearing material layer thickness. Under pulsing load with high amplitude, the 0.5-mm layer is susceptible to layer defects. Small porosities lead to early failure. The thicker layers show no such behavior. Here any small pores do not measurably reduce the fatigue strength and do not act as a crack starter. On very thin layers, the defects are relatively bigger, and so obviously their influence on failure increases. The best precaution against early fatigue failure is therefore to avoid extremely thin layers for practical applications.

Defects in a thin layer lead to a failure under pulsing load much earlier than on a thick layer.

The new fatigue tests, discussed in Section 6.2, showed no significant influence coming from surface roughness. In principle, rough surfaces result in higher lubrication temperatures and so reduce the fatigue capacity. But after some starts, the surface in the area of the smallest lubrication gap becomes smooth, and in practice there is an automatic curative effect.

More serious is the influence from defects on the surface. Scratches, especially across the running direction, can significantly reduce the fatigue resistance.

7.9 Lubrication Gap Temperature and Material Temperature

During our fatigue tests, the temperature measured in the lubrication gap was significantly lower as first calculated. The initial speculation that the variance was generated by unequal temperature distribution along the height of the lubrication gap, or was based on heat transfer between lubricant and material, proved false. The reason was the calculation procedure. A recalculation of all tested operational conditions with a 3-D calculation program, taking account of the pad deformation, confirmed all measured temperatures. The results from the program THRUST (7) deviated less than 2°C from the measured temperatures.

The temperature difference between top and bottom of a pad leads to a parabolic deformation. The smallest lubrication gap will become reduced further, but at the borders the gap increases significantly. Thus, the pad receives more cooling oil and the temperature drops. This effect has been incorporated into the 3-D calculation program and leads to proper conformity between calculation and test results.

Chapter 8

Preconditions for the Surface Lining of Plain Bearings

On early machines, white metal bearings were cast directly into the component (for example, main shaft bearings in an engine block). This lining process often gave poor results because only the static casting procedure was available, and the bulky complex component structure made uniform cooling impossible. Inevitably there were problems with frequent failures of the inadequate bearings. The next stage of development introduced the complete plain bearing as an exchangeable unit with housings inside the machine adapted to the new design. New lining procedures, such as centrifugal casting, could now be employed. The complete plain bearing, supplied ready for use, is today state of the art.

This development is a typical example of change under pressure of progress. Today, in many applications the service life of plain bearings exceeds the durability of the machine, and there is no need to change them. It is already clear that in future new design and fabrication, adaptations will be required.

8.1 Equipment for the Casting Shop

The fabrication of plain bearings requires substantial equipment such as a furnace for the heat treatment of the backings, shot blasting cabin for surface preparation with suitable grit, chemicals, tinning equipment, and a melting pot for the white metal. Tooling is needed for each of the different lining procedures involved.

8.1.1 Annealing Furnace

Electrically heated furnaces have a high efficiency and can be accurately controlled. In Section 3.4 the various heat treatments for plain bearings were described. The correct heat treatment is an important step on the way to good plain bearing quality. The closest control over heat treatment is achieved when the process is carried out by the bearing producer.

8.1.2 Grit Blasting Cabin

When a blasting cabin is used for rust removal or paint stripping, the circulating grit will become contaminated with these particles. If they come in contact with the bearing shell surface, a drastic reduction of bond quality between backing shell and bearing layer will follow. When the blasting treatment is subcontracted, the described risk is very high and out of control. Furthermore, there is the risk of new surface oxidation during transport from the grit blast contractor to the bearing producer. To get maximum control, the bearing producer should have his own blasting cabin exclusively for the treatment of bearing backings. The cabin has to be supplied with oil-free compressed air with a pressure of at least 5 bar, and these conditions must be checked regularly.

The blasting cabin must be located at the plain bearing shop and used exclusively for the backings.

The compressed air must be oil-free, and this has to be checked periodically together with the pressure.

8.1.3 Grit

Comprehensive tests have been made with different types and sizes of grit. Suitable types are cast iron grit, corundum, and steel cut wire, as shown in Figs. 8.1 to 8.3. For best performance together with long service life, cast iron grit with grain size 0.5–1.5 mm is preferred. With corundum grit and steel cut wire, similar quality results are possible, but the service life of these grits is shorter. When the grain size of the grit is reduced by wear to below 0.5 mm, it should be changed. Other grits, such as glass, are not suitable for treatment of bearing backings.

The grain size of the grit must be checked periodically.

8.1.4 Activator or Flux

To enable good tinning, the bond surface must be treated with an activating agent or flux.

There are procedures known to be in use in which the bearings are dipped in up to five different pickling baths, but this is far from state of the art. In addition to the huge space required, the variety of chemicals involved results in a serious quality risk.

Fig. 8.1 *Cast iron grit.*

Activators specially designed for the pre-treatment of bearing backings need just a few seconds to work by dipping or by brush application. For the dipping procedure, a plastic tank of activator is required. For brush application, a drip tray is required to catch the surplus activator. ROPTIN® (5) is an activator specially optimized for the dip tinning procedure. It has full action within seconds of contact with the backing, and the subsequent coating with tin is very effective. The liquid tin will even travel along the steel surface that remains out of the tin bath.

In contrast to pickling agents such as hydrochloric acid or sulphuric acid, there is no risk with ROPTIN® (5) of hydrogen absorption by the steel backing. This is a most important difference!

When dip or immersion tinning is not available, the wipe tinning procedure is an alternative that requires a different approach. The wipe tinning flux compound

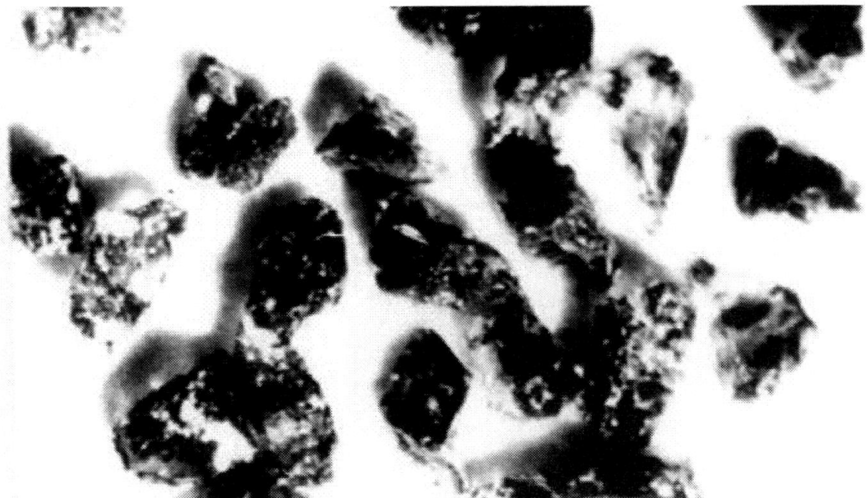

Fig. 8.2 Corundum grit.

Fig. 8.3 Steel cut wire grit.

contains pure tin, pickling solution, and flux materials, and during tinning there is always exposure to the atmosphere. To clear the contamination and facilitate the wipe tinning, a liquid soldering flux solution is used for activating the bearing surface.

Only such specialized activators are acceptable, as they do not allow absorption of hydrogen into the steel backing.

8.1.5 Equipment for Tinning

When the immersion tinning method is chosen, a heated tin bath will be needed. The bath temperature must be controlled automatically and set at 300°C. Regular calibration with an independent temperature-sensing device is advisable. An efficient fume extraction system has to be located above the tin bath to remove the aggressive vapor generated by dipping the activator-coated bearing shell.

When using the lead-free premium white metals, of course the bearing has to be tinned lead-free. The tin bath must not become contaminated, not even with residues from white metals, and therefore the melting out of old bearings in the tin bath is not acceptable. The impurities of the tin bath should be limited to 0.5% copper and 0.5% antimony. The permissible content of lead has to be limited to the same as for the white metal (i.e., 0.06%) or the lining will become contaminated and have a reduced melting point—see Section 4.3—and reduced bond strength.

When using lead-free white metal, the tinning has to be carried out with lead-free tin.

Periodic tin bath analysis is essential.

8.1.6 Melting Equipment for White Metals

Molten tin alloys are very corrosive to many other materials, and so the melting pot should be made from gray cast iron, which has good corrosion resistance. Other materials such as steel plate or even stainless steel have no resistance and therefore have low durability. Pitting corrosion occurs, and the molten white metal is contaminated by the dissolved material from the melting pot. To use overly large melting pots is not recommended due to rising energy costs and increased time and effort in handling the melt.

The white metal pot needs an adjustable automatic temperature control to hold the melt steady at the casting temperature recommended by the white metal supplier. The temperature measurement has to be made directly in the melt, and because of the corrosion problems the sensing element has to be protected. Gray cast iron tube is well proven for this purpose. Periodic temperature checks made with an independent device are advisable.

Steel melting pots are of no use for melting white metals due to danger of pitting corrosion!

8.2 Different Bonding Surface Designs

In practice there are various different designs for bonding surfaces, all intended to gain a good bond between backing and lining material.

8.2.1 Dovetail Grooves

The provision of dovetail grooves originated in the design of low-loaded plain bearings with cast iron backings. Because of the low bond strength achievable with this backing material, the additional mechanical locking design between lining material and backing was created. From the beginning, this design was a poor compromise. It is easy to see that a mechanical link between two materials, in place of a metallurgical bond, cannot guarantee long service life under dynamic load. But even when a metallurgical bond exists, the presence of dovetail grooves presents some drawbacks.

Plain bearings with dovetails inevitably have a lining of varying thickness. In practice, the variation is in the range of 2–5 mm and for old designs is even more.

For the same load, a thick layer suffers more elastic deformation than a thin layer. Therefore, along all the dovetail edges stress peaks occur. The combination of stress peaks with sharp edges at the same location is a likely crack starter and cause of failure, as every machinery trainee learns. Using the dovetail design ignores this basic rule, and damage can follow, as shown in Fig. 8.4.

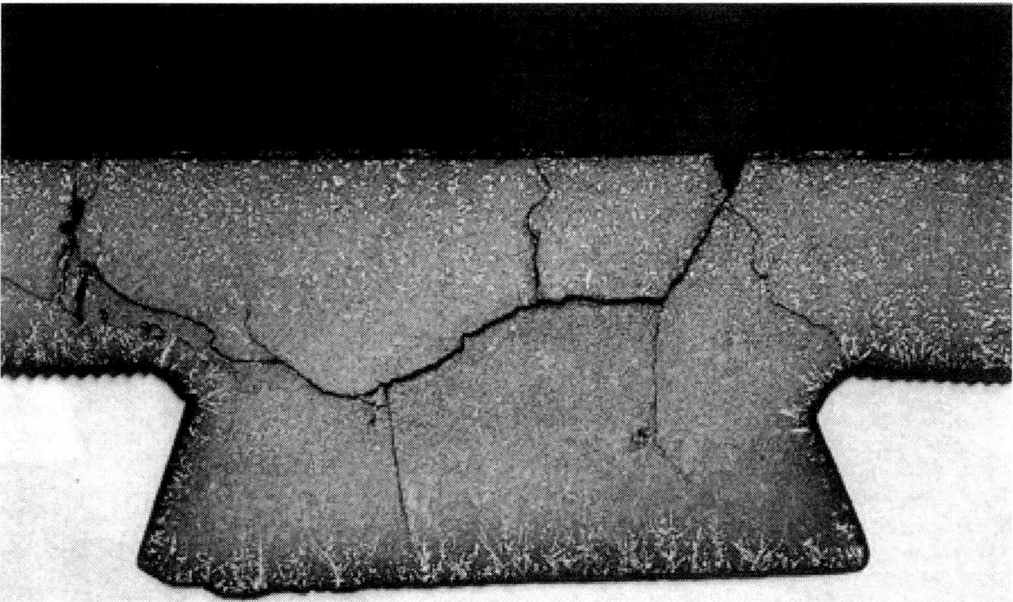

Fig. 8.4 Mechanical locking of lining material and backing, referred to as dovetails. Drawbacks: varying layer thickness, stress peaks, crack starters, early failure, and limited quality control.

After long running under radial load, the lining material becomes compacted in relation to the original thickness. With the dovetail design, the compaction happens unequally according to the layer thickness. Eventually the location of dovetails becomes clearly visible at the lining surface, and that means that unequal lubrication clearances exist along the bearing. On Fig. 8.5 these effects became clearly visible after deposition of oil carbon from high-temperature operation. The compacting effect produces further stresses at the sharp edges of the dovetails and leads to cracks and failure.

The dovetail groove recesses form a potential area for inclusions of air bubbles or dirt. That creates further quality risks. It is fatal that especially in this critical area the ultrasonic bond test is not effective, and therefore only 50% of a dovetail design bearing can be quality checked by ultrasonic bond testing. This fact cannot be altered by any amount of insistence on accurate quality control.

When lining dovetailed bearing backings by the static casting procedure, the grooves lying horizontally have to be interrupted at short distances so that the trapped air can escape from the groove. However, even with this precautionary action, the intended release of air bubbles cannot be fully achieved.

In spite of all the risks, this design has been in use for a long time and has also been applied, quite unnecessarily, to steel backings, despite the fact that excellent bond on steel surfaces is state of the art. The justification sometimes used is that, in case of damage, the dovetail design helps to center the rotor, but this has no standing against the well-known advantages of uniform thin layers on a modern plain bearing.

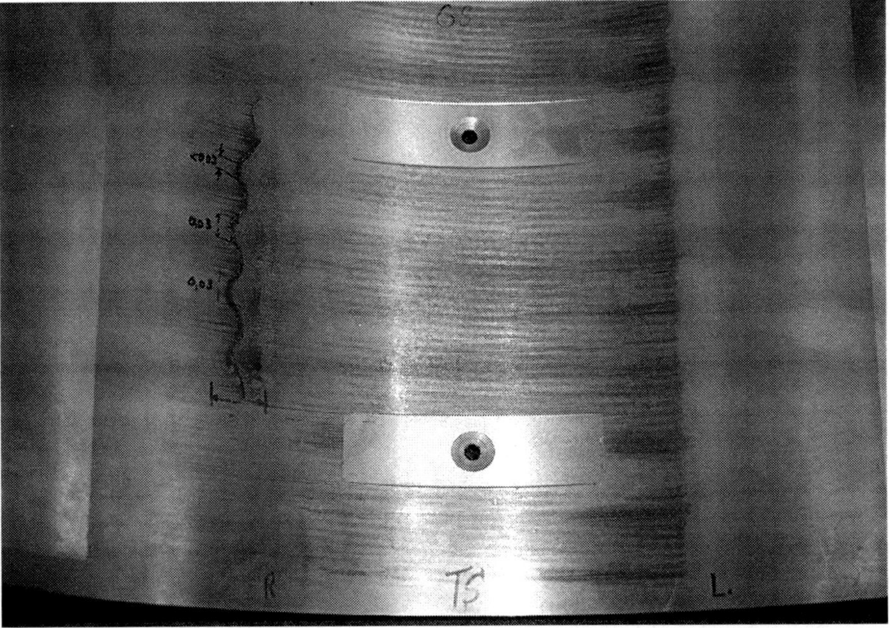

Fig. 8.5 Compaction in bands due to varying layer thickness. Here the effect became visible in the burnt oil film after overheating of the bearing.

The dovetail design is a demonstration that decisions made 100 years ago, and well proven under the conditions of those days, cannot be transferred to other applications without reviewing the change of conditions and the associated disadvantages. A design widespread in the industry takes a long time to revise. The temporary solution is not to accept the parameters, such as bearing diameter, used to make the decision for dovetails. Plain bearings with dovetails were already obsolete in the 20th Century. In the 21st Century they totally fail to satisfy the standards.

Design details that were, in former times, essential for the function have today become irrelevant due to completely changed basic conditions.

Plain bearings with dovetails are no longer state of the art!

8.2.2 Machined Bond Surfaces

Sometimes bond surfaces are found to have a wave pattern and in extreme cases sharp edges like a screw thread. This provision is intended to increase the bond surface, and therefore the bond strength, and initially the bond strength test seems to confirm this. However, looking more closely at the situation, we find that the bond strength tests always refer to the projected surface. So the increased surface is not taken into account, the relation between force and area is no longer correct, and the "better" result is seen to be false.

A lining on such a profiled surface very soon cracks under dynamic load. Cracks start from the summit of one peak and proceed to the next. Early failure of the total lining/backing compound follows, as shown in Fig. 8.6. Instead of an improvement, the design results in the opposite.

It must always be remembered that well-intentioned actions may produce the opposite result.

Bond surfaces with a wave or threaded profile result in early failure.

8.2.3 Smooth Bond Surface

Modern plain bearings have a smooth bond surface with uniform layer thickness. With proper preparation of the steel backing and correct lining of a uniform layer of bearing alloy, a high-quality bond with over 60% of the ultimate strength of the lining material can be obtained to give a safe, problem-free compound bearing.

An even bond surface is state of the art.

8.3 Preparation of the Backings

Plain bearings that are to be lined by a casting or soldering procedure require careful preparation of the backing metal. The process consists of heat treatment, surface blasting, flux activation, and tinning.

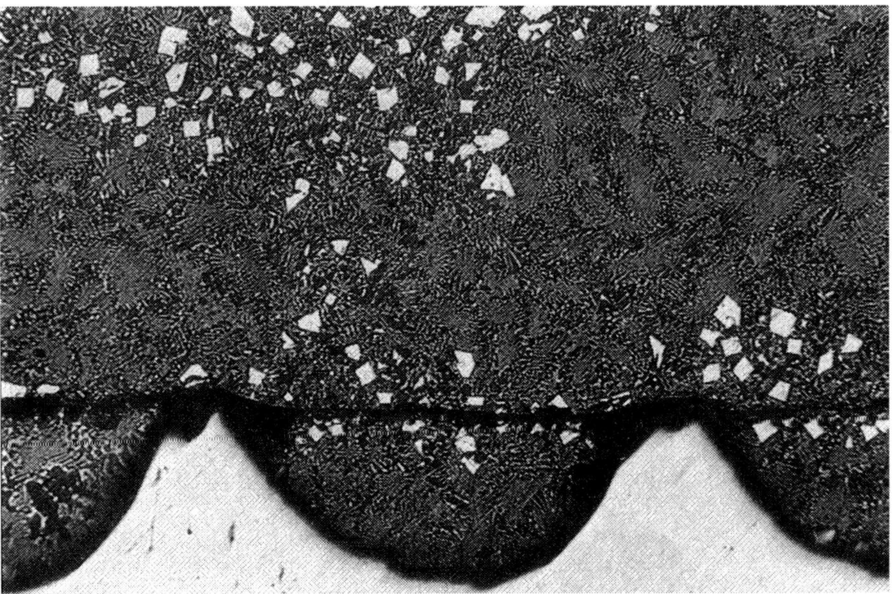

Fig. 8.6 *A screw-thread-style bond surface is no route to better bond. Under dynamic load, cracks start on the peaks and the bond fails.*

Occasionally, additional degreasing comes under discussion, especially for repair work.

Plain bearings arriving for repair work are dirty and greasy, and first of all have to be cleaned and degreased. The previously mentioned preparation steps can then start.

During these steps, no degreasing procedure should be repeated. Following blasting and activating, the surface is already metallically clean. An additional degreasing procedure would be detrimental because, after each degreasing, residues remain on the backing surface.

Degreasing activities immediately before tinning will damage the quality.

8.3.1 Heat Treatment

Correct heat treatment of the bearing backing is an important precondition for good bond quality. Time and again it has been observed that, on backings with repeated local bond failure during fabrication, the problem was solved after a proper heat treatment. It is a sure clue that the backing originally had no heat treatment or the wrong one (see Section 3.4).

The selection of the suitable heat treatment is very important.

Heat treatment is an important precondition for good bond quality.

8.3.2 Blasting Procedure

Before tinning and lining, the bond surface has to be prepared. The first preparation step is the blasting procedure. With this treatment, the surface becomes metallically clean and has an increased roughness. Immediately after blasting, the surface must be activated and tinned. Long intervals between blasting, activating, and tinning allow oxidation or even contamination with oily dust from the workshop atmosphere.

The blasting procedure has to be intensive and thorough. Because this is not easy to control, it has been found best for workers from the casting shop to carry out the blasting procedure themselves. The complete lining procedure is then the responsibility of one operator. The foundry men know from experience how to avoid extensive reworking. As a rough guideline, intensive blasting procedure means 50 minutes per square meter with a pressure of 5 bar.

Plain bearing producers that are not dealing with repair work will normally omit the blasting procedure. They prepare the new bearing bond surface by machining, the finishing cuts being made dry without any cutting oil.

Intensively blasted or dry machined surfaces lead to the same bond quality results.

8.3.3 Activating

It is not recommended to use a variety of fluxing agents. One special flux is absolutely sufficient. This avoids risks from deviations of composition of the different agents. The bond surface of the backing will be wetted with the flux agent. This can be done by dipping into a bath or by painting with the agent, and having the wetted surfaces is enough to ensure that the later tinning will be successful. When painting with flux, it is therefore important to do it accurately along edges and borders.

Only surfaces activated with a fluxing agent accept tinning.

Activating has to be carried out carefully, especially along edges and borders of the bond surface.

8.3.4 Tinning

8.3.4.1 Immersion Tinning

The tinning process can most effectively be carried out by immersion of the bearing backing in a tin bath of adequate size. Maintaining the required bath temperature of 300°C must be ensured by permanent temperature control. With a higher tin bath temperature, there is a risk of unacceptable oxidation of the tinning after removal from the bath. If the bath temperature is too low, the tinning process fails.

Besides correct bath temperature, a sufficient immersion time is important. Too short a soaking time leads to insufficient backing temperature and makes proper tinning impossible. Conversely, a long immersion time for the backing in a tin bath is not detrimental to the tinning process.

There is a risk of underestimating the soaking time, especially on thick-walled backings. The time is sufficient only when the backing is completely heated through to a temperature of 280–300°C. In practice, the problem is that the backing surface very quickly reaches the bath temperature, and a surface temperature check will always suggest that the process is complete.

That might be wrong. To check for correct heating right through the backing, it can be taken from the tin bath and the surface temperature measured immediately and again after 3–5 minutes. A significantly lower temperature on the second measurement is a clear indication that the center of the backing is not yet hot enough. The immersion time was too short and must be extended.

The immersion time required depends on the size of the tin bath and the mass of the backing, so universal figures cannot be given. Each casting shop has to make tests with different backings, based on the above procedure. The collected data will indicate adequate immersion times for the different masses of backings and reduce the need for frequent temperature checks.

Complete heating through of the backing is a precondition for successful lining by casting and a good bond between lining material and backing.

With an inadequate immersion time of the backing in the tin bath, there is a risk of process failure. A long immersion time increases the quality safety factor.

Multiple tinning operations do not improve the tinning quality but also do no harm. Therefore, after tinning, a bearing backing can be allowed to cool down, then at a later time heated up ready for casting by reimmersion in the tin bath.

Multiple tinning procedures do not improve the tinning quality.

The volume of tin in a tinning bath can be reduced by using displacement devices. When a tinning bath is not needed for a while, and the heating is switched off, an insulated cover to seal the bath will conserve the heat of the molten tin and save energy for the subsequent heating up.

8.3.4.2 Wipe Tinning

When no tinning bath is available, the tinning has to be done manually with a specialized tinning flux. A paste containing pure tin powder and fluxing agent is applied to the bond surface of the backing and heated by gas torch flame on the outside of the backing until the paste melts and the tinning takes place. Uniform heating is important, and local overheating is a quality risk. These conditions are not easy to fulfill when the backing is big and has various thicknesses. On big backings, the heating time is longer and there is an increased risk of the paste drying out too soon. Following the tinning process, the surface has to be thoroughly washed with clean water to remove flux residues and to avoid unacceptable oxidation of the tinning. Before casting the white metal, the shell has to be reheated to casting temperature.

For successful wipe tinning, an experienced operator is essential. The immersion tinning process carries much less risk thanks to the quasi-automatic uniform heating and tinning during immersion and the absence of flux residues. On removing the backing from immersion tinning, it will be at the correct temperature and can go directly to the casting operation.

Immersion tinning and wipe tinning both can lead to the same good bond quality; however, the wipe tinning process contains many more quality risks.

8.4 Correct Use of the Molten Lining Material

The white metal is heated in the melting pot, becomes molten, and must then be further heated and held accurately at the recommended casting temperature.

Directly before taking molten white metal from the melting pot for casting, there are essential procedures to follow.

The slag or dross has to be skimmed from the surface, and the melt should be thoroughly stirred with the ladle. When ladling the metal from the pot, it should not be taken from the surface but from deep in the center. In cases where the melting pot has a bottom outlet, the melt should be mixed with a stirrer before casting. The shaft and paddles of the stirrer must be from cast iron to avoid the intensive pitting corrosion that can affect steel, as mentioned in Section 8.1 (see Fig. 8.7).

If no further requirement for molten metal is foreseen, the heating may be switched off and the melt allowed to solidify in the open pot.

The melt must not be held at a reduced temperature for any length of time. The consequence of holding the melt between liquidus and solidus temperatures is that copper (Cu) crystals will separate out and grow under these conditions. They become concentrated on the bottom of the melting pot as a thick viscous layer, shown in Fig. 8.8. Several problems will result. The separated copper will not go back into solution when heating up the melt the next time, so this quantity

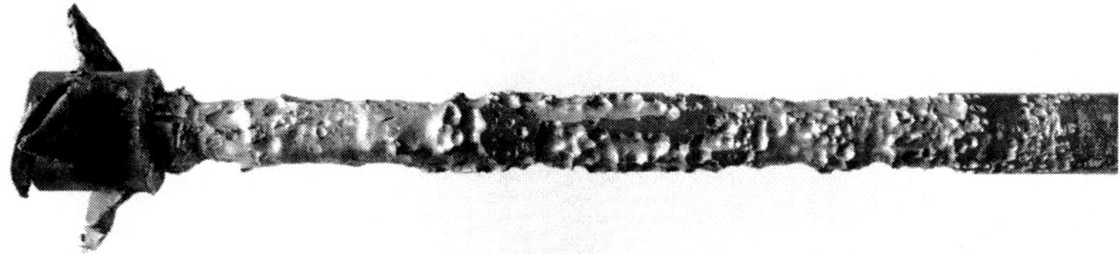

Fig. 8.7 A cautionary example. Shaft of stirrer and paddle made from X5CrNi18-10. The melt induces intensive pitting corrosion. The same occurs on steel melting pots. To solve the problem, the use of cast iron is recommended for melting pot, stirrer shaft, and paddle.

Fig. 8.8 *CuSn layer from the bottom of the melting pot.*

is missing from the alloy specification. When such separated copper is cast into a bearing, these large, hard crystals will be torn out of the surface during machining and damage the bearing.

Whenever the melting pot is empty, it should be checked for a copper layer at the bottom, and any found should be removed.

Chapter 9

The Lining of Plain Bearings

Several different lining procedures have been developed over the years. At first, all white metal bearings were made by the static casting procedure. This method was followed by other methods such as centrifugal casting, soldering, flame spraying, and more. This diversity was not automatically a continuous improvement of the lining results. The different procedures have specific assets and drawbacks. Often, a particular lining procedure is requested based on the belief that it will supply the best quality. However, each selected procedure is linked to conditions that influence the result. If the conditions are not met then the consequence is bad quality.

To order a special lining procedure does not automatically lead to good quality. The conditions of the selected procedure must be achievable.

For all lining procedures, the focus of attention should be the solidification of the molten lining material. The solidification must always start at the bond interface. A second solidification front beginning at the lining surface must be avoided. Ignoring these conditions will result in shrink holes or complete failure.

It is also important for the solidification to be rapid and evenly distributed along the bearing to generate a uniform and fine crystallization. To meet this condition, the backing has to be intensively and evenly cooled immediately after casting the lining material. For backings with varying thickness, the uniformity of cooling cannot be fully guaranteed, and where the backing wall is very thick, a sufficiently fast cooling may not be possible.

For a high-quality cast lining, the solidification must always start from the bond interface and has to progress uniformly and rapidly.

For casting of linings with tin and lead-based alloys, the backing temperature is specified as 280°C, whereas the bearing alloy will be at a considerably higher temperature (e.g., between 440°C and 540°C, depending on alloy composition). The figure of 280°C takes account of the fact that on one hand a higher backing temperature results in better bond strength and on the other hand high backing temperature increases risk of oxidation of the tinned surface. The temperature of the poured alloy will drop to a level between the alloy and backing temperatures. This temperature clearly has to be above the liquidus of the alloy to allow for a controlled solidification by water cooling.

9.1 Static Casting

Static casting is the oldest procedure for lining plain bearings. Figure 9.1 shows the procedure in use in 1969. Today it is less important because the achievable quality is falling behind the other lining procedures.

Fig. 9.1 *Superlative static casting at Th. Goldschmidt AG, 1969. Bearing of a cement mill with inner diameter of 5200 mm.*

For static casting, journal bearings will be mounted, with axis vertical, on a casting plate after heating to the specified temperature. A corresponding preheated former will be brought into position so that between former and backing is a uniform gap. The molten alloy will be poured into the space between backing and former. The cooling will be by water sprays to the outside of the backing, starting from the bottom of the bearing and proceeding upward. Solidification of the cast alloy starts at the bottom, proceeding upward, and at the same time starts at the bond interface, proceeding inward. To achieve both solidification directions continuously, without interruption, the water spray has to follow just behind the upward solidification and must never get ahead. At the same time, a wire will be used vertically to gently agitate the liquid alloy. This procedure releases air bubbles trapped in the melt while allowing the solidification level to be monitored.

The correct direction of solidification from the backing toward the former can be realized as long as the former temperature is higher than the backing temperature. If these conditions are not met, the consequences are shrink holes and bond failure.

At the beginning of the static casting process, the bottom of the bearing is always at a higher temperature than the top. Consequently, the static casting procedure can never give totally uniform crystallization and bond strength along the full length.

The static casting procedure brings further quality risks: very often, the formers are steel with thin wall thickness and consequently the casting process becomes uncontrollable. While a 50-mm-thick backing needs intensive water cooling, a former of 5 mm thickness simultaneously cools much faster by atmospheric cooling influence alone.

The fact is that the requirement for solidification to start from the bonding surface cannot always be met. Many attempts have been made to solve the problem by cooling the backing from outside with water and heating the former from inside by flame simultaneously. These actions go in the right direction, but in practice the problem is merely modified. By the nonuniform distribution of the flame along the former surface, some areas become overheated and other areas lose heat. With each change of flame position the heat distribution is changed due to the former's thin wall thickness. Uniformity as specified for the process is impossible in practice. Improved uniformity is only attainable by using formers of greater wall thickness.

Often one former will be used for different-length bearings of the same diameter, and so the former is sometimes much longer than the bearing. The projecting part cools down very quickly in the air and toward the end of the static casting process adds to the problem of temperature imbalance. The backing is already colder than at the beginning of the process, and the former is too cold due to its excess length. Shrinkage pores and bond failure are the result.

Similar temperature problems may occur at the bottom of the static cast bearing. Although at the start of the casting process the backing and former both have the correct temperature, the casting plate allows intensive heat loss. Even a well-preheated casting plate very quickly loses temperature, much quicker than the backing and the former. So at the base of the bearing, porosity and bond problems may also be caused by temperature fluctuation.

Static casting of cylindrical bearings cannot produce uniform quality. There are too many influences on temperature deviation and hence a quality reduction. Process control is only partially possible. The realization of a reasonably stable process requires comprehensive practical experience, and thus the process acquires some mystique. Experienced workers achieve a more or less acceptable result, but knowledge of the respective conditions remains imprecise. On this basis, reproducible work is uncertain.

The required experience has been almost lost in the past few decades. Just a small number of workers remain familiar with this process, and they no longer apply it in the daily routine because static casting is rarely specified, except for repair work to old bearings.

More often the static casting procedure will be used for lining flat surfaces such as pads of thrust bearings. In these cases it is usual for the bond surface to be set horizontally and fitted with a perimeter wall made from steel plate, sealed to the pad. After pouring the metal into this "basin," the pad must be cooled from the underside. This is often accomplished by lowering the pads partway into a water bath.

Unfortunately, problems with nonuniform temperature distribution also affect pads. The edges cool quicker than the center, and the alloy solidifies there while

Fig. 9.2 *Static casting on a thrust bearing pad. The nonuniform solidification between center and border is clearly visible.*

it is still liquid in the center of the pad (see Fig. 9.2). Shrinkage porosity is the result. If, during casting, the pad temperature is already too low, bond failure occurs along the borders. With increasing pad size, the nonuniformity in cooling increases, and more compensating actions have to be taken. So for big pads, a concentrated water jet must be directed to the center of the underside, and the periphery needs to be heated at the same time. Even when these actions are successful in gaining temperature uniformity, there remains for big pads a further problem: the cooling speed is too slow to form a fine and homogeneous crystallization structure.

For example, in some hydro power plants there are pads with thickness of up to 400 mm. Clearly, in such cases a uniform cooling down in a short time is impossible.

The limit of pad thickness for acceptable quality has not been determined because it depends on so many influences. Also of great influence is the white metal. High-quality alloys, as for example TEGO® V738 (5) and TEGOSTAR™ (5), are in general only suitable for static casting under very restricted parameters.

The static casting procedure carries the risks of uneven temperature distribution, of variable solidification direction, and of falling below the acceptable solidification speed. These risks cannot be controlled, and the static casting quality results are lower compared with other lining procedures, especially for bigger parts.

9.2 Centrifugal Casting

The centrifugal casting procedure is ideally suited for lining the bore of rotation-symmetric bearing backings. The heated backing will be clamped between the endplates of the casting machine (shown in Fig. 9.3) and then rotated. The white metal will be poured in, and is uniformly distributed over the total bond surface by the centrifugal force. Water cooling is then applied to the outside of the rotating backing until the solidification of the white metal is complete.

Generally, it is accepted that the centrifugal casting procedure provides the best quality. Theoretically, that is absolutely right because the cooling is in principle very uniform and the direction of solidification is clearly defined. Furthermore, this procedure is very economical thanks to the short preparation and lining times. Nevertheless, in practice the procedure can also bring problems.

The backing must be preheated evenly to 280°C. When fixing the backing into the casting machine, there must be no loss of heat into the clamping plates. On small machines this risk can be offset by intensive preheating of the end plates. For bigger machines an insulating seal in the contact area is necessary to sufficiently limit the heat-flow problem.

The unavoidable small temperature losses at the bearing ends can be compensated for by pouring the liquid white metal mainly to those areas. This ideal can most easily be achieved when the pouring is done by hand ladle. If the liquid metal is poured via a casting channel, then selective distribution is only possible if the casting channel is mobile or there are two outlets set for the bearing length. Fixed casting channels have the drawback of pouring the liquid metal into one place and causing local overheating of the tinning. The liquid metal flows from the middle to the axial bearing ends and cools down on its way. At the ends, the temperatures of both poured metal and bearing back can be too low. Porosity and bond failure can follow. There may also be bond failure in the overheated area where the pouring channel was fixed.

The pouring channel itself can also result in dramatic heat loss. The white metal temperature can drop along the channel by up to 100°C. To compensate, the pouring channels must be adequately preheated and the temperature of the liquid alloy checked at the bearing end of the channel.

When pouring the liquid metal into the rotating backing shell, turbulence must be avoided. Ideal for this is an entry point about 30° before bottom dead center.

***Fig. 9.3** Centrifugal casting machine.*

For good bond strength, the liquid white metal must be at the proper temperature at the moment of contact with the backing shell. The temperature in the melting pot cannot be assumed to be the contact temperature.

The reference for the required rotation speed is the minimum diameter formed by the liquid white metal. Because this diameter cannot be measured exactly in practice, the curves in Fig. 9.4 refer to the inner backing diameter and are divided into thin layer (<10 mm) and thick layer. The curves relate to a centrifugal force of approximately 20 *g*.

When lining bearing halves, two halves are assembled together with separating shims in the joint faces. The inner overhang of the shims should be not more than half the lining thickness, otherwise they may cause intensive local turbulence. The ideal material for the separating shims is stainless steel because it will not bond with the white metal. The white metal covering the projecting shims will split open itself during cooling down.

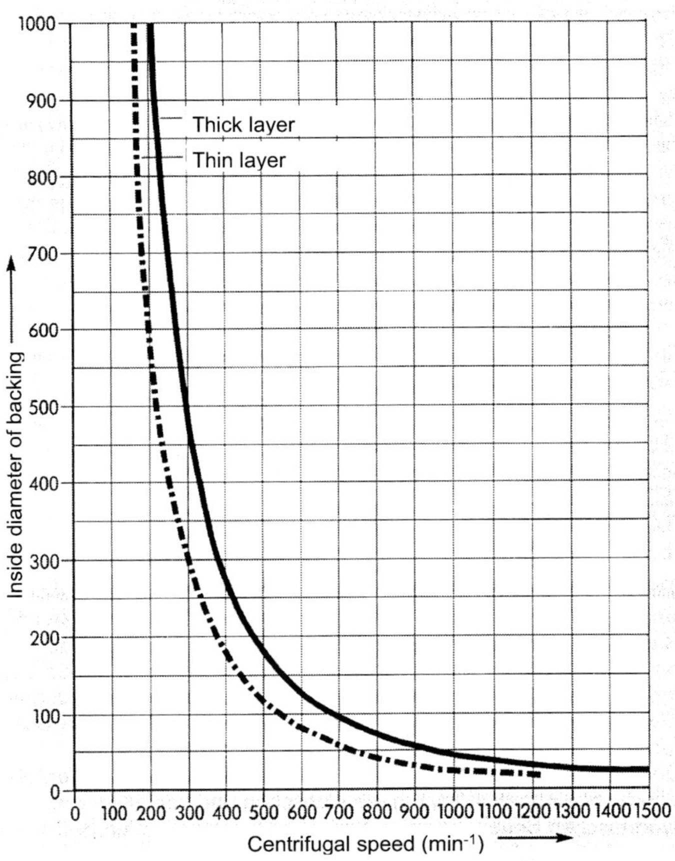

Fig. 9.4 Curves for selecting the centrifugal speed. The curves refer to the backing inside diameter and cover normal casting thickness (<10 mm) and thick castings.

Sometimes the cast surface is of variable thickness with tongue-shaped irregularities in the direction of rotation. If this appearance is found on every casting, then the centrifugal machine is unbalanced, and abnormal noise and vibration can usually be observed during the casting procedure.

If only some castings are affected, then another possibility may be inadequate and slow cooling. One part of the melt is already solidified along the bond surface, and the remaining melt sets later as a separate layer. Inhomogeneity is the result.

While the variable thickness caused by an unbalanced machine is a problem only when there is insufficient thickness for machining, the second influence of poor cooling generally leads to a reduction of quality.

Intensive cooling will always be needed. For this, the quantity of water is less important than the water pressure and the water jets formed by the spray nozzles. This powerful contact between water and the backing surface is crucial, otherwise a steam layer will be formed at the backing surface and the heat transfer much reduced.

In practice, many nozzles are positioned along the bearing length, usually in two or three rows. The nozzles must have individual stop valves to enable an adaptation of the cooling length to the bearing length. This avoids the unnecessary cooling of the clamping discs. Tilting nozzles are always helpful when the bearing backing has thickness variation, as the cooling can be concentrated to the thick walled areas. The temperature of the cooling water certainly has an influence, but not so much as the previous mentioned parameters, the more so as the water temperature cannot easily be reduced.

The water quantity is not alone in controlling cooling. A powerful impact of the water jet on the backing surface is especially important.

It must now be mentioned that the centrifugal casting procedure is not appropriate for every size of bearing. The limitation is not just the dimensions of the centrifugal casting machine. In fact, the necessary quick cooling will be harder to achieve as the backing size, and especially the thickness, increases. Even with optimization of all the previously mentioned influences, the physical limit for a homogeneous solidification of the white metal will finally be exceeded.

With increasing mass of the backing, the cooling time is extended to the detriment of the lining material quality.

The limiting factors between good homogeneity with a fine crystallization structure and inhomogeneity with coarse crystals cannot be clearly determined.

For the first time, investigations have been carried out on large radial tilting-pad journal bearings: the starting point for fabrication of these bearings was lining of a backing ring by centrifugal casting. The inspected bearing ring had a weight of 5000 kg and a wall thickness of 250 mm. The centrifugal casting was done on a machine with optimized cooling nozzles, with a cooling water quantity of 25 m³/hour at 6 bar water pressure. After lining, the pads were cut from the ring and the remaining areas between the pads were available for detailed inspection. The test results were alarmingly poor. The bad result is by no means unique to this bearing. Very often bearings with comparable and even bigger dimensions are fabricated, but usually as journal bearings where a sample of material for destructive testing is not available without sacrificing the total bearing. Comprehensive testing is therefore impossible, and the problems cannot be detected.

Extremely large centrifugally cast bearings always have a low lining quality. This has not been detected in the past because the quality control procedures applied did not address the problem.

As just mentioned, material specimens were taken from the surplus area between the pads. The lining material was removed in 1-mm steps and an analysis made after each step. There was severe material segregation, clearly due to the cooling rate being too slow. The different elements of the alloy were unevenly distributed through the lining thickness. The biggest deviation from the specification involved antimony and copper; (see Figs. 9.5 and 9.6). Figure 9.7 shows the distribution of all elements, and clearly antimony and copper have large deviations, whereas the elements cadmium, nickel, and arsenic show good homogeneity. This stems from the fact that in the solidification sequence, copper and antimony solidify first and move in the remaining liquid of the melt. The inhomogeneity happens through the lining thickness as well as along the periphery.

For another ring of similar size, the temperature of the cooling water was reduced to near freezing point. This action gave some indications of a step in the right direction, but the material analyses deviated even further from the

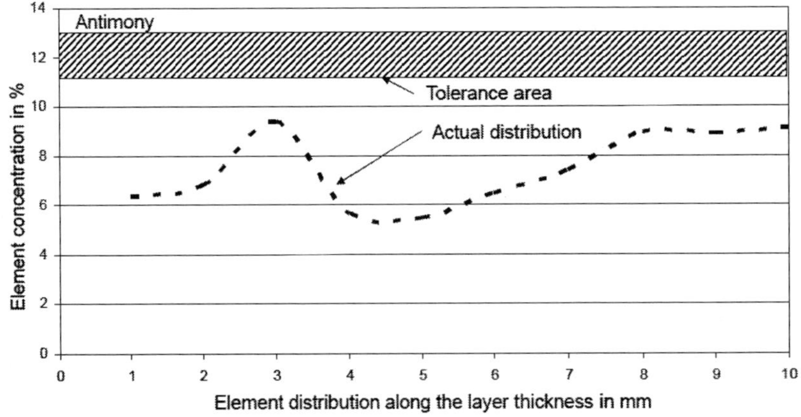

Fig. 9.5 *Deviation of antimony concentration through the lining thickness of a plain bearing with extremely thick backing.*

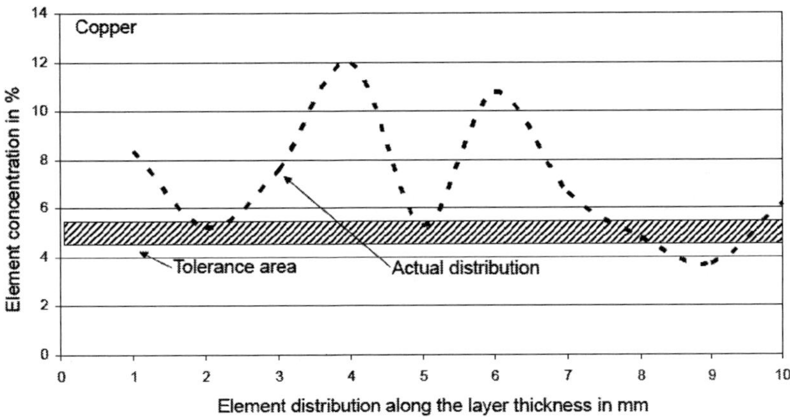

Fig. 9.6 *Deviation of copper concentration through the lining thickness of a plain bearing with extremely thick backing.*

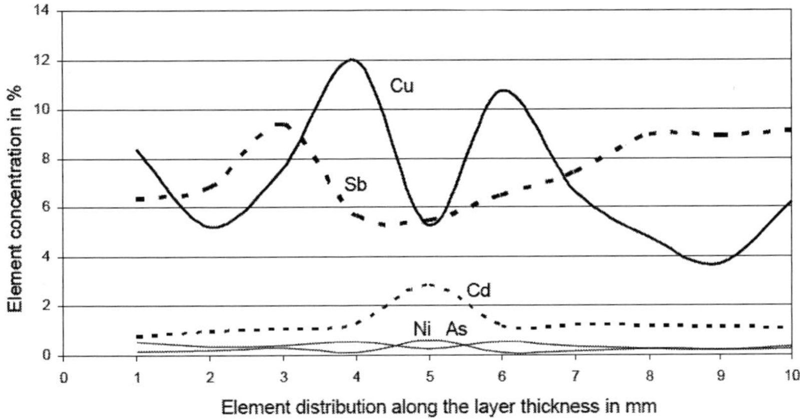

Fig. 9.7 *Deviation of element concentration through the lining thickness of a plain bearing with very thick backing. All elements of the alloy are shown.*

tolerances. The results show clearly that the centrifugal casting procedure comes to its limit when a large wall thickness is involved. For the described case, the limits were clearly exceeded.

This experience teaches us that it is not helpful to insist on centrifugal casting in all cases. For bearings with massive shells, such as the described example, selection of a different lining procedure is necessary for high quality and a homogeneous lining.

The shell thickness at which the quality of a centrifugal cast lining becomes critical cannot be quantified, because intensity of water cooling is different for each casting machine. However, even when the conditions are optimized, alternative lining procedures should be considered for backing wall thickness above approximately 150 mm.

These limits of backing wall thickness for centrifugal casting are no more than a rough guideline, because the casting machines in use worldwide are individually designed and not based on any standard. Cooling intensity and corresponding cooling times are therefore different and not comparable, and so the point of transition to unacceptable quality varies from machine to machine. Because the tendency to bigger plain bearings with thick backings continues, especially in the power plant industry, it is assumed that the limit of good lining quality there was long ago exceeded. In the future, reasonable care should be taken when selecting a suitable lining procedure for thick-walled plain bearings.

In the future, it should no longer be assumed that centrifugal casting is always the best lining procedure. For bearings with massive backing wall thickness, other lining procedures lead to better results.

9.3 Lining by Soldering

When lining plain bearings with a soldering technique, sticks of white metal are melted by burner flame and the molten material applied to the tinned backing surface. The procedure has been successfully used for decades and, when correctly done, has reached the same or better quality as centrifugal casting.

There are persistent and unjustified criticisms of this lining method, perhaps because the soldering method is often used to repair the faults created by other lining procedures. Frequently, these repairs have little success. When a multitude of bond defects are found in a plain bearing lining, this is a clear indication of low bond quality of the total lining. When, in such a case, local repair work is made by soldering, there is risk of failure in adjacent areas due to the local heat input to poor quality bond. In the end, such a bearing has endless repair spots, even visible on the running surface, and finally is rejected by quality inspection. This scenario has nothing to do with the basic quality of the soldering procedure. The problems come from impossible repair work on a poorly lined bearing. It is by such circumstances that the soldering procedure has been discredited and is rejected, with the explanation that it always gives poor quality. Certainly the soldering procedure can fail, as can all other lining procedures, but professional soldering lining on a properly tinned backing surface results in the same or even better quality than expected from centrifugal lining. The big advantage of the soldering procedure is the absence of negative influences from the backing thickness. This means there are no technical restrictions for the soldering procedure, and the only disadvantage is the amount of time needed.

Professionally soldered lining gives the same high level of quality as expected from centrifugal casting.

The soldering procedure can be applied without restriction.

It is often said that not all types of white metal can be successfully applied by soldering. This is incorrect.

In the early days of white metal bearings, only lead-based alloys were soldered. From the technical point of view, the lead-based alloys are very easy to work, but there are big ecological objections regarding the intensive fumes of lead, cadmium, and arsenic when melted by direct flame contact. An effective air extraction system is a must; however, in the European Union, soldering with cadmium-containing alloys is forbidden from 2012.

With the widespread change from lead- to tin-based alloys, it was found that the melting and flow behavior of the two groups is different. Variations in lining quality occurred and gave rise to comments that some alloys are suitable for the soldering lining procedure and others not. Particularly interesting are the often conflicting statements. For example, while one company suggests that soldering with TEGO® V738 (5) is impossible, another company does the work very successfully. Clearly the parameters for handling the various alloys are different, and while they remain unknown, the lining work is done intuitively with variable success.

Every tin- or lead-based alloy can be used for the soldering lining procedure with consistent results of a high quality level.

The author has studied the differences, systematically evaluated them, and developed a procedure to give good results for both lead and tin alloys. The procedure is as follows:

For the solder lining procedure, only burners with sieve nozzles should be used. These give a flame that is relatively softer and of increased section. This aids uniform melting of the white metal stick and reduces the flame pressure on the molten lining surface.

The soldering procedure operator must be constantly aware that he is running two operation sequences at the same time and that they are independent of each other.

One operation sequence is focused on the correct melting of the lining surface, whether it is tin on a backing shell or an existing white metal layer. As soon as a molten spot is created, the burner must be moved in the direction of the required metal build up. The molten spot consequently follows the burner, and a condition is established in which the burner height and speed of travel remain constant. These conditions must be maintained throughout the total soldering process.

The other operation involves the proper melting of the solder stick with the flame and adding the result to the molten spot on the lining surface. To achieve this, the melting rate of the stick has to be matched to the burner's rate of travel. The melting rate can be coordinated by positioning the stick nearer to or farther from the center of flame. Here are the recommendations for the different white metals. Lead alloys or tin alloys with high tin content (such as $SnSb_8Cu_4$) need less melting energy. Using these alloys, the stick has to be in a position away from the flame center, and in practice, the stick is very near to the lining surface.

Tin alloys with lower tin content—such as TEGO® V738 (5) or TEGOSTAR™ (5)—need more melting energy, which means that a stick position near the flame center will be necessary (see Fig. 9.8). This is a problem area for inexperienced operators who, finding that the stick does not melt fast enough, instead of moving the stick position nearer to the flame center, instinctively move the burner nearer to the stick. By doing this the constant conditions of the first described operation sequence are lost. By moving the burner closer to the lining surface, the melting spot becomes overheated and porosity and bond defects follow.

Practical experience repeatedly shows that, even when aware of both the independent operation sequences, the worker still reacts in the wrong way as described. To have the correct coordination requires lengthy training. It is very important to have expert supervision during training, because not every beginner is able to recognize their own incorrect operation coordination. It should also be remembered that during the lining soldering process, the backing must not be heated above 80°C. Application of cooling or processing pauses may help to stay under this limit.

Porosity in a solder-lined layer always points to overheating during the lining process caused either by reducing the burner distance or by overheating of the backing.

A correct working position is very important to achieve long periods of working without fatigue. The operator should sit in a position adapted to the orientation of the surface to be lined so that supporting the forearms on the knees is possible. This way a consistent movement coordinated between burner and stick can be made without tremor. The movement will be controlled by knees and wrists, with the weight of burner and stick supported by the legs; see Fig. 9.9.

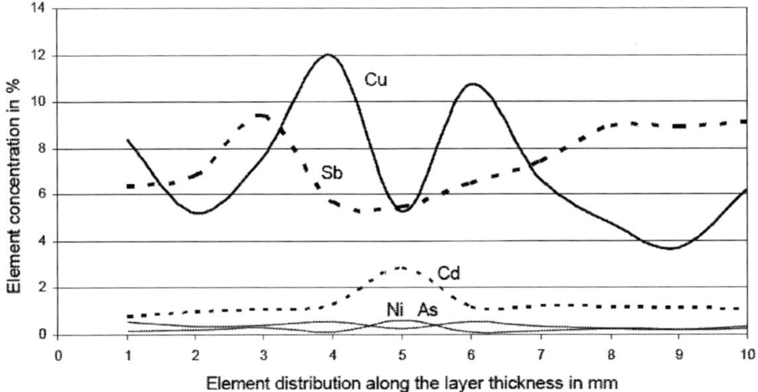

Fig. 9.8 The soldering lining procedure. This photo shows the soldering stick in position near the flame center. Below that is the molten spot where the liquid white metal must fall.

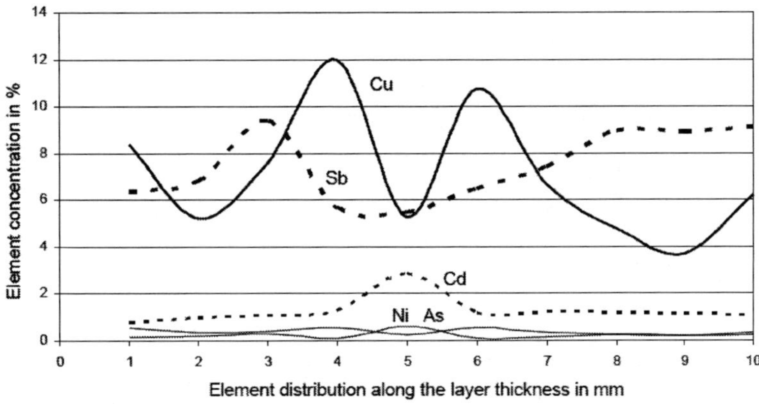

Fig. 9.9 Supporting the forearms on the knees allows uniform and tireless working.

During lining by soldering, two operation processes are running in parallel at the same time and are independent of each other. The proper coordination of these processes requires training under an experienced instructor.

Alloys containing lead, cadmium, or arsenic should not be lined by the soldering procedure due to the health risk from fumes during the process.

9.4 Thermal Spraying

Occasionally, bearings are lined by flame spraying. This procedure is well proven for metals with higher melting points, such as copper alloys. Alloys based on lead and tin have much lower melting points and therefore react differently when applied using the flame-spraying procedure. The chemical composition of the white metal lining differs from the original specification following flame spraying. In particular, cadmium, arsenic, and zinc evaporate. This reduces the strength of the lining because the elements for quality improvement are no longer present in sufficient quantities, and some essential crystallization cannot occur. A similar problem happens with other spraying methods such as plasma spray. During the flame-spraying procedure, the transit time of successive particles is long compared with their solidification time. Consequently, semi-molten particles surrounded by oxides hit the surface, burst, and only partially fuse together. The structure formed by these particles is lamellar, and oxides are embedded. Lines of porosity will be formed parallel to the running surface in thin layers; see Fig. 9.10. The formation of bigger cavities is also possible. Flame-sprayed white metal layers are porous.

The structures visible on all the unetched photos of flame-sprayed white metal layers shown here are not crystals but cavities. The lining has an anisotropic nature, with technical properties greatly reduced in the direction of thickness. Measured in this direction, the hardness is reduced to 30% of the normal value.

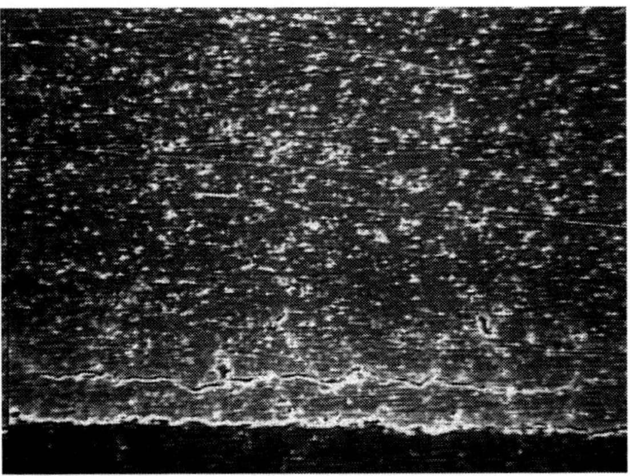

Fig. 9.10 Section with horizontal crack along a chain of pores, unetched.

Fig. 9.11 Cast white metal as a comparison, etched.

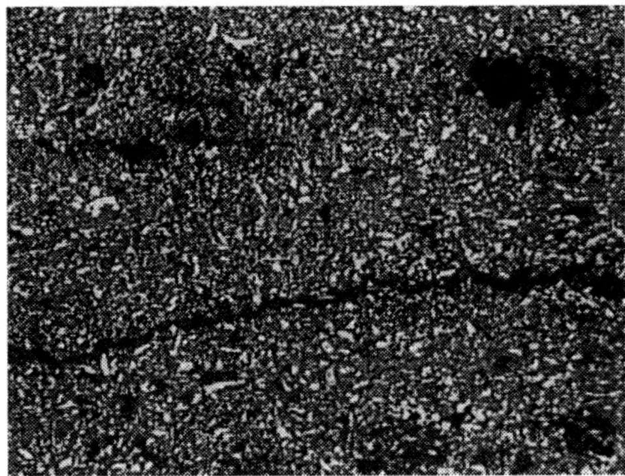

Fig. 9.12 Flame-sprayed white metal, etched.

Generally, for white metals the hardness test is not a good indicator of the overall quality, but in this special case of 30% hardness loss due to the lining procedure, it is obvious that there are such changes to the material due to the flame-spraying procedure that it is not comparable with the original alloy. This problem becomes clearly visible on the photos. The crystallization structure is not comparable with the structure of the cast alloy (see Figs. 9.11 and 9.12). On the specimen, the white metal can be split horizontally with a fingernail, as shown in Fig. 9.13. The sprayed layer can be broken by hand and falls into parallel pieces; see Fig. 9.14.

Bearings lined by flame spraying give no signs of problems when first supplied. However, when they are carrying load, it is not long before dramatic damage occurs. The lamellar lining cannot transport the shear forces, cracks start, and the lamellar layers fail. The layers can carry perpendicular loads in compression but will not cope with tensile forces. Parallel to the running surface, the material can withstand tensile loads but has no capacity for compression and shear. The material clearly demonstrates anisotropic characteristics comparable with the behavior of a stack of paper; see Fig. 9.15.

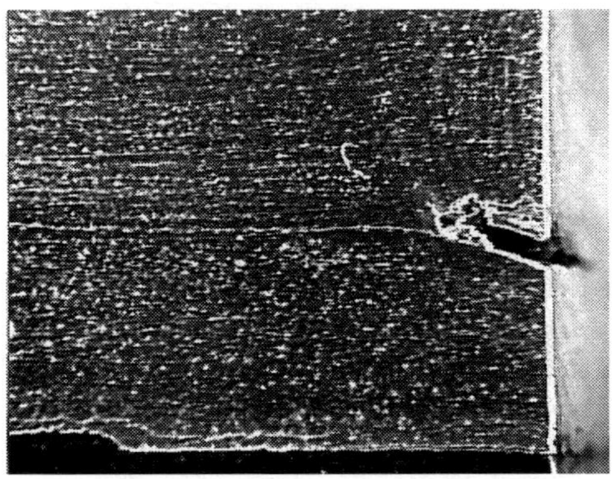

Fig. 9.13 *Flame-sprayed lining with layers of porosity chains. On the right, split just with a fingernail, unetched.*

Fig. 9.14 *Flame-sprayed lining can be broken by hand into layers.*

An example from practice:

When a blister became visible on the lining surface of a bearing, it was at first assumed that hydrogen diffusion was the reason, as shown in Fig. 9.16. But the newly-lined bearing had a very old backing, and so the hydrogen release effect was ruled out. Further investigation finally discovered a flame-sprayed layer with a thin surface lamination separating off, obviously initiated by some small negative local pressure; see Fig. 9.17.

The next example also shows the inadequate quality of flame-sprayed white metal bearings. In Figs. 9.18 and 9.19 we see a burst white metal layer in an oil pocket area. The thin layer of the oil pocket could not withstand the thermal periphery stresses. The lining delaminated across the width of the oil pocket. Both ends of the lamination flexed, and the lining in the oil pocket area became a slender bending and buckling beam. This beam finally became unstable and, subjected to low pressure in the pocket area, permanently buckled.

Figure 9.20 shows the pad of a journal bearing from a hydropower machine. The top layer of the lining is flaking off.

The service life of a flame-sprayed highly loaded plain bearing is limited. The damages described occur very early, after only a few load cycles, and ultimately lead to the damage characteristic of dynamic overload. The reason for the damage is the anisotropic material behavior, with insufficient material strength caused by the lining procedure. Low load capacity and reduced service life of flame-sprayed plain bearings leads to the strong recommendation that this procedure not be considered for white metal lining.

The white metal material properties listed in the manufacturer's technical data sheet are not correct when the alloy is used for lining by the flame-spraying procedure.

(a) Surface can withstand compression

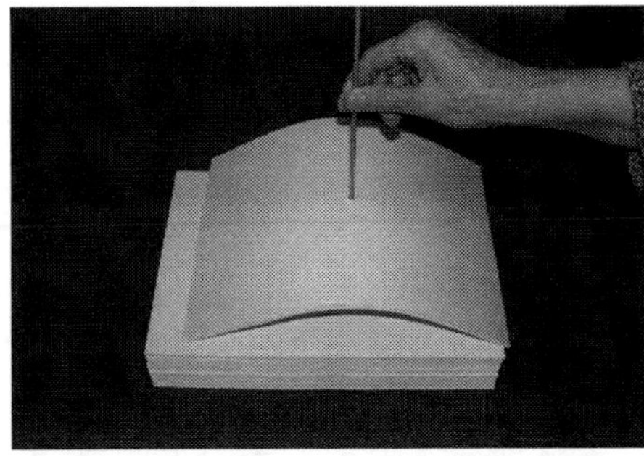

(b) Surface cannot withstand tension

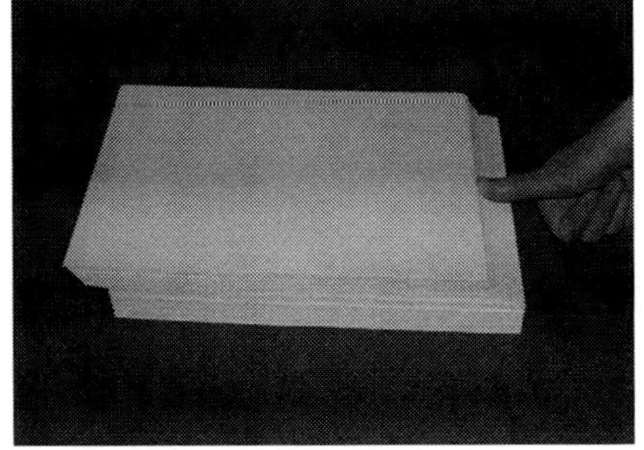

(c) No shear resistance

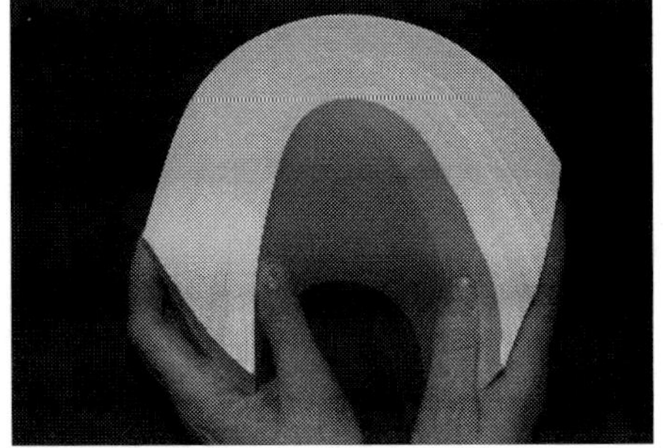

(d) No bending load capacity

Fig. 9.15 *The flame-sprayed layer has behavior similar to a stack of paper.*

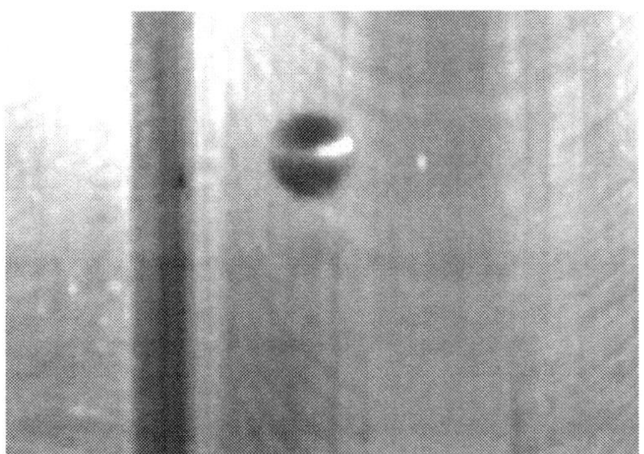

Fig. 9.16 *Blistered surface is often a result of hydrogen problems, but investigations showed that here a thin surface layer of a flame-sprayed lining has been separated by negative local pressure.*

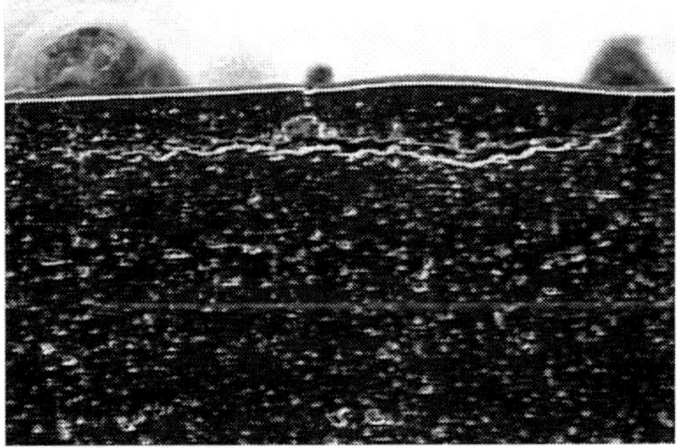

Fig. 9.17 *Material separation near the surface. It is the damage from Fig. 9.16.*

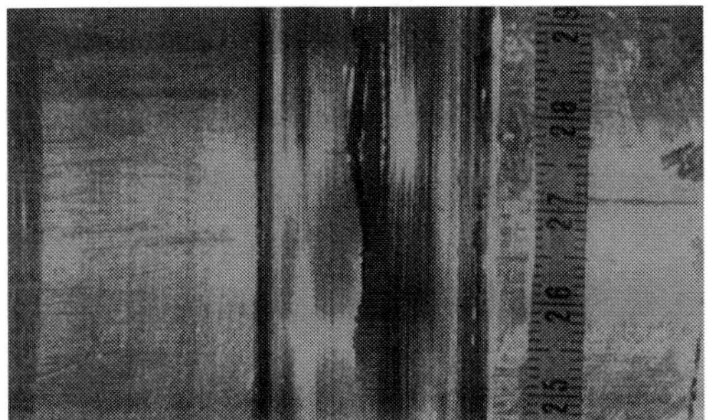

Fig. 9.18 *Damage to a flame-sprayed white metal layer in the area of an oil pocket.*

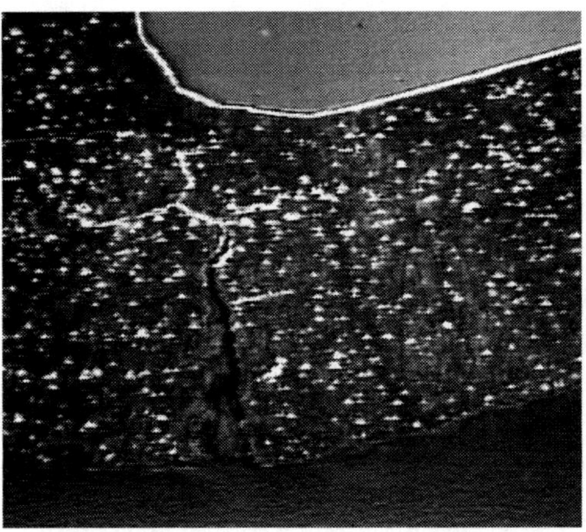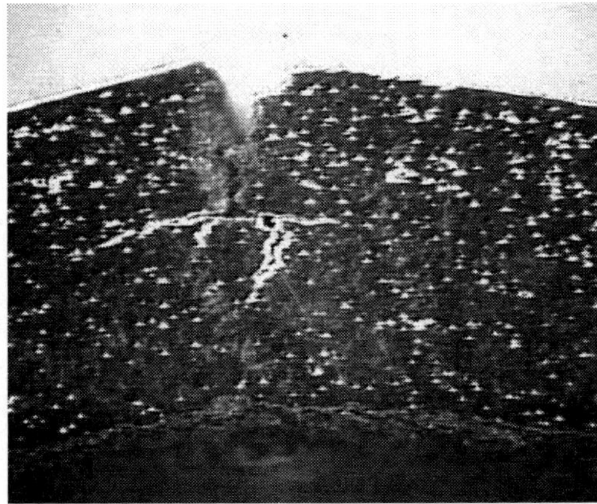

Fig. 9.19 *Delamination of lining, right and left borders of the oil pocket form pin joints, and the slender "beam" buckles under peripheral stress.*

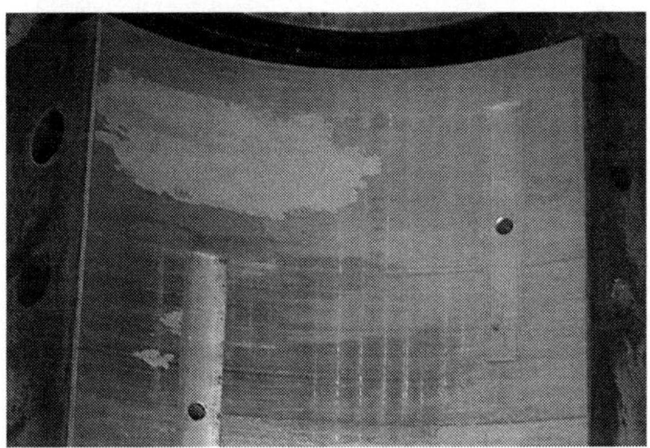

Figure 9.20 *shows the pad of a journal bearing from a hydropower machine. The top layer of the lining is flaking off.*

Flame spraying of white metals results in uncontrolled melting and solidification and is not an acceptable procedure for high-grade plain bearings.

The documented material properties are not valid when lining white metals by thermal spraying.

9.5 Cold-Spraying Technology

When lining by cold spraying, the material is supplied as powder and is carried by an ultrasonically accelerated process gas. The particles are projected onto the substrate with high velocity and adhere to form a layer. The kinetic energy at the moment of impact is not sufficient for controlled melting of the white metal particles, so deficiencies exist similar to those of the previously described flame-spraying procedure.

The cold-spraying technology in current use does not allow complete and controlled melting of white metal particles, and therefore the procedure is not suitable for lining of plain bearings with this material.

9.6 WIG Technology

The WIG technology is a welding procedure that has different names in different countries: WIG (Wolfram Inert Gas welding), TIG (Tungsten Inert Gas welding), GTAW (Gas Tungsten Arc Welding).

For lining plain bearings, it is not economical due to the attainable welding speed and layer thickness. The biggest restriction comes from the white metal. Just a few tin alloys are available as wire, and their properties relate more to solder materials than to white metals. To produce high-grade white metal wire is not practicable. When WIG lining with tin alloys, the fusion penetration is uncontrollable. The procedure could be used for spot repair work, but in most cases the required white metal quality is not available in wire form. WIG welded linings have finer crystallization, which in principle is an advantage, but it means that the structure of the repaired spot is different from the remaining lining, and that is not acceptable.

9.7 The Different Lining Procedures and their Advantages and Disadvantages

The lining procedures are very different and have individual limitations. Further limitations result from the backing geometry, as shown in Table 9.1. In most cases, lining is a manual operation, and the result therefore depends on the worker's

Table 9.1 Overview of Different Lining Procedures in Use for Plain Bearings

	Advantages	Disadvantages
Static casting	• For all backing geometries	• Non-uniform solidification • Big grain of crystallization • Quality depends on the worker's qualification
Centrifugal casting	• Homogeneous and find grain crystallization • Uniform bond • Economic procedure	• Only for rotation-symmetric backings • Increasing backing thickness results in bigger and inhomogeneous grain
Lining by soldering	• For all backing geometries • Fine and homogeneous grain • Uniform bond	• Quality depends on worker's qualification • Time and cost intensive
Thermal spraying	• Automatic procedure	• Unregulated solidification • Banding with linear porosity and inclusions of oxides • Porous and anisotropic layer with insufficient material properties
WIG-technology	• Automatic procedure • Fine grain	• Use of wire limits selection of lining material • Small welding beat, thin layer, slow, uneconomic • Uncontrolled melting of backing surface

experience and accuracy. Consistent quality cannot be guaranteed. That the quality control staff cannot observe and deal with the problems just mentioned is the biggest concern. It could be addressed by continuous destructive tests, but that is not practicable.

The ideal solution would be a new lining procedure that allows constant high-level quality without any restrictions from the process or from the geometry.

9.8 Plain Bearing Lining Technology of the Future

The lining procedures in current use have some limitations, and during the last 15 years the industry searched with great effort for improved qualities of lining materials compared with TEGOSTAR™ (5), but without success.

AlSn alloys and bronzes are both able to carry high loads, but they can fail spontaneously and can damage the shaft or thrust face.

Tin alloys have the reputation of carrying less load than Al or Cu alloys, but they have the unbeatable advantage of slow failure progress, in which the sliding partner remains undamaged. It is, therefore, logical to start R&D activities for improved plain bearings based on SnSbCu alloys. Because long service life depends significantly on the stability of the lining geometry, for future development of durable tin-based plain bearing materials, the zinc hardening, as with TEGOSTAR™ (5), provides the only chance to combine long life with increased load.

The mission, therefore, is to improve the technical properties of the zinc-hardened alloy SnSbCu while simultaneously improving the environmental considerations, to simplify the lining process and especially to achieve repeatable quality. In the future it will no longer be acceptable to line plain bearings worldwide with the same white metal, and when checking the structure, mostly in the case of bearing damage, to find a wide range of different crystallization because of various casting conditions worldwide.

During the last ten years, no significant progress has been made. The developments move in a circle, because under the present marginal conditions, the development potential is almost exhausted.

For many years, the author has been conscious of the situation and is convinced that the precondition for leverage is to change the present limiting conditions. In this way, new horizons will be created and the way will be open to explore the potential for completely new developments.

In this respect, it is particularly helpful to view the problems from an alternate perspective and to have the courage to question the usual actions.

Under the present marginal conditions, the development potential is almost exhausted. Therefore, the conditions have to be changed.

Try to assess and justify the usual procedures and when in doubt, question them!

This may seem to be a tall order, but it is actually not unrealistic. Here the road to success will be explained and shown to lead to a completely new phase of plain bearing development.

We started with an intensive investigation focusing on bond, crystal refining, and material characteristics.

9.8.1 Bond

The different influences on the bond strength have already been discussed here. Quality of bond is primarily defined by the quality of the lining material because the tensile strength of the lining material sets the maximum limit. Bond strength is reduced by temperature deviation during the lining process and also by poor preparation of the backing or a bad tinning procedure. On bearing failures in which lining material becomes detached, the layman wrongly blames weakness of bond, but the culprit is a completely different mechanism.

For various reasons the bond quality always becomes the center of quality discussions. That seems logical because, besides the bond test, there is in practice no applicable quality test procedure. It is remarkable that the destructive bond test is, in practice, hardly ever used and that the ultrasonic bond test gives no information about the bond strength. In practice, therefore, in most cases reduced bond strength remains undetected. Despite this, the ultrasonic bond test enjoys universal acceptance.

For the tinning procedure, some fluxes produce intensively aggressive fumes when the bearing is dipped into the tin bath. For environmental reasons, it is urgent to find a remedy. At the moment we have the situation that without flux there is no tinning, and without tinning there is no compound (bi-metal) bearing.

The question arises whether there is any alternative method to achieve a high-quality compound bearing without flux and even without tinning.

Research based on avoiding the need for flux activating and tinning has finally resulted in the discovery of a superior alternative.

Is it possible to get a good compound (lined) bearing without using flux and without tinning the backing?

9.8.2 Crystal Refining

The group of elements that will refine the crystal structure of tin alloys is well known. Some of these elements are toxic and others not economic. Most commonly used is arsenic. Selenium and chromium have also been found to work, but because the alloying procedure is complex and the refining result is no better, they have no industrial application. With the development of HOYT® 11Z3 (5) and TEGOSTAR™ (5), refining with silver increasingly came into use.

The metallurgical refining of the crystal structure is not difficult, but of concern is the fact that the refining effect is strongly influenced by the cooling conditions during the casting process. Put simply, the metallurgical refining effect can become completely lost when bad backing geometry or bad casting conditions apply.

The question arises as to whether there is any alternative to crystal refining that is independent of the backing geometry and gives consistently good results.

By questioning the practice of crystal refining with additional alloying elements, a procedure has finally been found that always gives a fine structure, independent of existing backing geometry.

Is it possible to get uniform crystal refining of the lining material irrespective of the backing geometry?

9.8.3 Material Characteristics

Chapter 4 discussed the elements required for achieving improvement of material characteristics. The number of appropriate elements is finite, and most of them are toxic. It has long been known that, for example, a higher copper content improves the properties of the lining material. But with the lining technologies in current use, a copper content above 6% is unacceptable because the compound fails.

The idea occurs that perhaps there is a way of producing a compound (bi-metal) bearing without restrictions in the composition of the lining material.

This calls the existing lining procedures into question, generally, and justifiably so. Considering the many insufficiencies and restrictions at present, it is hard to believe that in the future, plain bearing casting will remain unchanged from the way it has been done the last 100 years. Future increasing pressure on process stability cannot be met with the existing lining processes and monitoring methods.

Is it possible to line plain bearing backings without restrictions imposed by the lining process?

9.8.4 Laser Lining Process

Detailed investigation of alternative lining processes finally focused attention on laser technology. It is not new and is already established in many areas of industry. Where applied it always leads to:

- ultra-fine crystallization

- maximum homogeneity

- minimized weld penetration

- repeatable results

These are exactly the properties we want for the plain bearing linings, but first there was the difficulty that existing experience with laser lining technology was restricted to applications with chromium, cobalt, wolfram, and similar materials with high melting points. Second, the usual linings with these materials are much thinner than needed for plain bearing lining. There was no experience with lasers of depositing a large thickness of low-melting-point material. Therefore, we started to research the subject.

At the start of our research, there was no really suitable material available. Therefore, the first trials were made with solder wire and then later followed by tests with $SnSb_8Cu_4$ wire, although these materials possess intensive creep behavior and are not qualified for high-loaded plain bearing lining. However, concerning the lining process, very soon the general advantages of laser lining described earlier were proven to apply to tin-based alloys.

For laser lining with tin-based alloys, the flux activating and the tinning process are no longer needed, and the attained bond strength is always on a level significantly higher than that of cast linings and with a result variation of surprisingly narrow spread.

The crystal structure is many times finer than that seen with cast or soldered linings, and the crystals have a completely homogeneous distribution, irrespective of the backing's geometry; see Figs. 9.21 and 9.22.

Laser lining with copper alloys is also successful, and so there are for the future no restrictions on the composition of alloys. The way is open for further improvement of plain bearing alloys without process restrictions, and we arrive at a new beginning of plain bearing metallurgy with extensive potential for optimization.

All the desired ideal features are achievable with the laser lining technology, and the repeatability and homogeneity are guaranteed to be at the highest level.

After the initial euphoria came the sometimes painful detailed optimization work to achieve the laser-lined results economically. To be competitive with the casting procedure, we set a target of lining 1 m² with 3-mm thickness within two hours. At first glance this may seem to be slow, because the casting procedure takes a much shorter time. However, it has to be considered that the casting process also requires the preliminary steps of fluxing, heating, and tinning and then, after casting, time for cooling. These time-consuming steps do not apply to laser lining.

To fulfill the demands of a high lining speed of 2 m/min, a high-capacity material feed is needed. Multiple problems occurred with this. With high lining speed, the required energy input per unit length increases significantly, and consequently so does the laser power needed. In the early stages, the laser power requirement was underestimated, and for a long period the tests were made with

 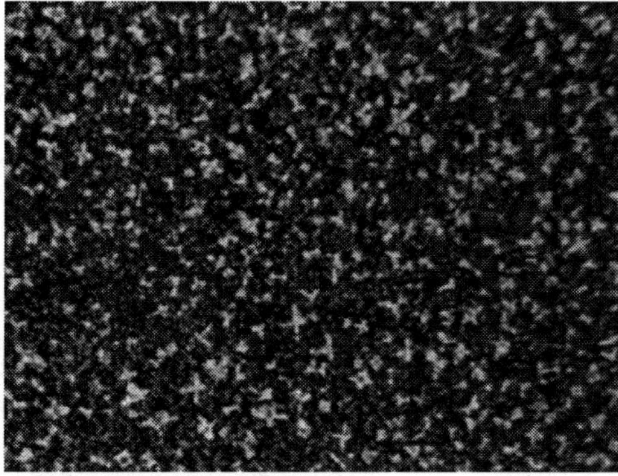

Fig. 9.21 Left, cast lining. Right, laser lined. Both linings are made with the same specification of lining material, and the photo magnification is identical.

a far too small 4-kW laser. The result was insufficient bond. After changing to a 10-kW laser, the problem was solved, but another problem arose.

Backings of low mass lined at high speed and with high-energy input naturally heat up rapidly, and need temperature-controlled cooling. For practical use, it is clearly more economic to reduce lining speed and energy for small backings. There is a wide linear relationship between laser power and speed, and material feeding. That makes modifying the parameters easy. At present, the tests are made with a round or oval laser spot, but it is already possible to modify the laser spot to a rectangular shape with adequate powder feeding. This modification should improve the economy for plain bearing lining.

The operational tests clearly showed problems with feeding the lining material as wire. Not every white metal composition can be made into wire form, and this limitation would be an unwelcome restriction on the material optimization. In addition, tin alloy wire gives irregular results when laser welded. The wire center melts differently to the wire surface, producing variable structures along the layer section, which means reduced homogeneity. Finally, the feeding of sufficient wire and correct distribution is problematic, especially when using modified laser-spot geometry. To eliminate all these restrictions, we have changed from wire to powder feed for all further process optimization work. Practically every alloy can be produced as powder, and powder particles melt consistently at each laser-spot position. Figure 9.23 shows powder feeding into the laser spot.

By using powder feedstock, all restrictions on the lining process are eliminated.

Currently we feed the alloy powder with an argon gas shield. Under high laser power and lining speed, the standard gas protection is not sufficient, and pores form in the lining. An additional argon gas supply was found necessary to protect the complete welding area. Formation of pores can also be avoided by using a helium atmosphere, but this choice of gas is less economical. Future work

Fig. 9.22 *Laser-lined surface. The individual weld beads are clearly visible on the unmachined surface.*

will be concentrated on optimization of the lining material. Here there is the possibility for unprecedented advances in quality. As an example: the standard alloy $SnSb_8Cu_4$ has an ultimate tensile strength of 77 N/mm², and the first white metal developed for laser lining, LACOMET® (5), has 125 N/mm², already better than $AlSn_{40}$. Figures 9.24 and 9.25 show white metal test specimens.

Admittedly the tensile strength does not represent the full material properties or the material specific loss of strength with higher temperature. Nevertheless, it can already be stated that, with the change from casting to laser technology, the tin alloys can match the properties of AlSn alloys without losing the tin alloys' advantages under damage conditions. This is the ideal starting point for further development without restrictions from material and process.

With the change of method from casting to laser lining, tin alloys do not lose their unique failure advantage during plain bearing damages, but they gain the loading level of AlSn linings.

When material properties greatly increase, it is no longer sufficient just to compare the conventional properties. New test procedures representing the more intensive loading of the plain bearing have to be found and used—see Chapters 5 and 6. At this point, the loop will be closed between new demands, new lining procedures, new lining materials, and the specially adapted performance test procedures.

Conventional documented material properties are not representative of the material when lined onto a backing and, in any case, for laser-lined layers the conventional data cannot be produced. The layers produced by laser lining technology are too thin to make the 10- or 20-mm-thick test pieces for the conventional compressive, rotating bending, or impact bending tests.

Fig. 9.23 *Laser lining with white metal. The powder feeding into the laser spot is clearly shown. The laser beam itself is invisible.*

Fig. 9.24 *White metal test specimen compressed 50%. On the right is cast material with typical cracks along the surface. On the left is lasered material of the same analysis. The specimen has been deformed 50% without forming any cracks.*

In the future, as well as tensile tests, data for fatigue resistance under different conditions will be recorded. It will be beneficial to develop suitable standards to make the fatigue test results comparable.

For material data of lasered linings to be comparable, new test standards will be needed.

The advantages of the laser lining procedure clearly show the great potential for plain bearings. There is the opportunity to combine the good damage behavior of tin alloys with the high loading capacity of bronzes. The material family LACOMET® (5) aims at doing this already. The brand label hints at the procedure (Laser-Compound-Metallurgy) and refers also to the characteristic lining material (Laser-Coating-Metals).

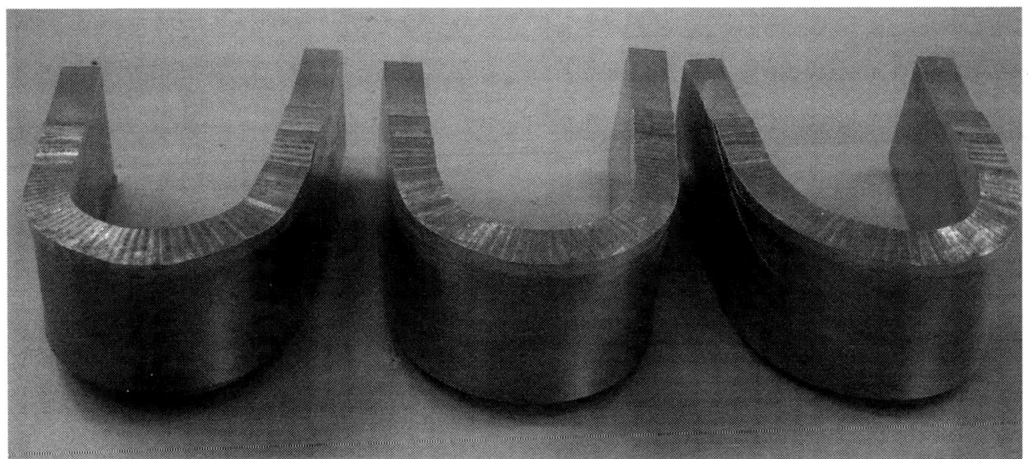

Fig. 9.25 *Compound of white metal on steel made with laser technology, bent 180° without cracking or bond failure. Such a test result is not possible with a cast layer.*

9.8.5 Summary of the Advantages Given by Laser Lining Technology

Fluxing and tinning of backing shell are eliminated: less working steps, environmental problems solved, costs for tin bath and energy saved.

Significant improvement of bond strength with high uniformity:

- repeatable quality of high level.

Absolutely homogeneous and ultra-fine crystallization structure:

- improved and repeatable material properties.

Crystallization structure independent of backing geometry:

- consistent high quality.

Material thickness needs just to allow for final machining:

- saving of material, less scrap.

Total result repeatable by defined parameters for automatic lining process.

Lining material composition is not restricted by the lining process:

- unlimited optimization of lining material is available.

Combination of alloys with nonmetallic composites becomes possible:

- new applications possible.

In the long term, no foundry equipment will be needed:

- the lining operation will be just a working step of the mechanical processing.

Chapter 10

The Quality Standards for Plain Bearings

In producing plain bearings, as for most jobs, many details have to be considered so that in the end a high quality can be obtained. To begin with, the manufacturer needs experienced employees. But that alone is not sufficient to ensure uniform quality of a high level. A specification of required quality and applied quality control are both needed. Between the two, a well-balanced relationship is necessary, and this in practice is often missing. For the customer, it is easy to ask for the maximum possible quality, but can that be achieved? Is it necessary and reasonable? In a single case, the supplier may agree to a customer's excessive quality request to win the order, but in the long run the customer always pays for requesting excessive quality.

Economic quality is that which is just sufficient for the particular application.

Quality control usually concentrates on the finished bearing. To keep the scrap rate low, the plain bearing producer should extend the quality control to the complete fabrication process by monitoring every step.

10.1 Plain Bearing Producers' Process Qualification

In spite of process monitoring, experience shows that the risk remains of unnoticed deviation of results in the course of time. In such cases, the typical assertion is that everything has been done as always, but suddenly the result was different. Clearly that cannot be true. The fact is that not all conditions were originally identified and included in the schedule of observations.

Here are two outstanding examples:

If there is a reduction in cooling water pressure during centrifugal casting, or the nozzles are wrongly adjusted or even blocked, the cooling conditions are compromised and the quality deteriorates. This has been detailed in Section 9.2.

If the monitoring of water pressure and nozzles is not included in the operation instruction, then it can be assumed that periodic control will not take place and deviations will not be recorded. If there is suddenly a bad result, the search for the reason begins, everything is checked, but the cause will not be identifiable.

A similar situation may arise with temperature control. The temperatures for tin bath and white metal pot are set to the nominal value and are then automatically controlled. However, is the indicated value identical with the real bath temperature? To identify defects of the temperature gauge, periodic checks with an independent thermometer should be made directly in the molten metal. Comparison of this result with the display of the permanent automatic measurement gives direct information about unacceptable deviations. This procedure must be included in the operation schedule, or it will not be considered.

This book contains many hints on various topics from material selection, through heat treatment, preparation and tinning, to the different lining procedures. When all the hints are considered with the specification and quality control, then most preconditions for process stability and high quality are covered. Problems are mostly based on influences that remain unidentified and therefore unconsidered.

Finally, there are the conditions to be met that are individual and different for each casting shop; for example, the holding time of the backings in the tin bath for complete heating (see Section 8.3.4). The centrifugal casting procedures are also individual and different due to the absence of defined standards.

The quality of the tinning and lining work can only be checked in practice by a destructive bond test. In a production operation, that is in most cases not possible. The practical solution is, therefore, a qualification procedure of the casting shop followed by documentation of all relevant parameters.

The logical proceeding is as follows:

A backing with geometry similar to the usual production plain bearings is run through all working steps. All preparation conditions, and especially the casting conditions, have to be documented in detail. Subsequently, specimens for destructive bond testing (ref. ISO 4386-2) will be taken from the cylindrical bearing center area and from the cylinder ends. The tested bond strength should not significantly differ between center and end positions, and all should be at a high level. More details about this are in Section 10.3.2.1.

The white metal layer from the test piece should be metallurgically examined for homogeneity. Inhomogeneity is an indication of insufficient cooling conditions. Low bond strength is a general hint that process temperature was too low or temperature losses were too great. If the test result is not satisfactory, the total fabrication process must be repeated with modified parameters until the result is good. The documented parameters from the successful process will be used as references for future fabrication. For the qualification process, the backing may be re-used by welding in steel plates to cover the holes left by cut-out specimens, but the next specimens must not be taken from a position near such closed holes due to locally unrepresentative cooling conditions.

Obviously a plain bearing of 100-mm bore has different parameters than a 600-mm bearing. When a bearing shop fabricates many sizes of bearings, the qualification process has to be repeated for each size, and all appropriate parameters have to be documented. It is advisable to rerun the qualification process from time to time. To fabricate small test samples during the normal production process is not helpful because these small parts have different parameters, and their test results are not relevant to the fabricated bearings.

For the process qualification, backings have to be used of similar quality, design, and dimensions to the intended production bearings.

10.2 Quality of the Plain Bearing Alloys

The alloy composition has to be specified before an accurate analysis can be done. It is widely accepted that the quality is sufficiently described by the specification and the related analysis certificate. Little known is the fact that observing the rules of composition formulation is an important precondition for long life of a plain bearing.

10.2.1 Specification of a Plain Bearing Alloy

Most materials have a brand name (e.g., TEGOTENAX (5), MB90, and ASTM Grade 2). Many standard alloys are characterized by the chemical formula, such as $SnSb_8Cu_4$ denoting majority Sn (tin) with 8% Sb (antimony) and 4% Cu (copper).

The chemical formula lists the most important elements of the alloy but does not specify tolerances, small quantities of further elements, and permissible levels of impurities. This crucial information gives the correct specification. Unfortunately, most international standards say the element with the biggest quantity has to be declared basically as balance, and only the other alloying elements are specified with tolerances. This ignores simple mathematical rules to get the alloy composition to appear as a nominal 100%.

Required elements have to be clearly differentiated from the impurity elements by having maximum and minimum limits. Permissible impurities should always be specified with "max" or "<" or "≤." For example: Pb < 0.06% or Fe max. 0.03%. That is necessary to avoid confusion between permissible impurity and necessary alloying element, but in practice, confusion exists with some standard alloys for which they ignore these simple rules. This often leads in practice to quality problems and endless discussions where no difficulty needed to exist.

Until now all the relevant DIN and ISO standards were guilty of inconsistencies as discussed. At the request of the author, the standards are now finally being revised, and we have the promise of consistency for the future.

A simple but important rule: material specifications have to clearly differentiate between alloying elements and impurities.

Alloys declared as "lead-free" nevertheless need the limit of permissible lead impurity to be specified. When defining the impurity limit, the achievable measurement accuracy has to be considered. For practical use, Pb < 0.06% is a well-proven limit. The measurement techniques cover this range, and the concentration is below the level of detrimental influence on the material properties.

For environmental reasons, it is not acceptable to eliminate the documentation of permissible impurities and, in any case, that would mean the quality is not sufficiently defined for technical purposes. To clarify:

When a specification defines only the constituent elements that have to be analyzed and documented, any elements not specified, not analyzed, and not documented nevertheless may be included in the alloy at an unacceptable concentration. In the case of lead, this could become critical for quality and the environment.

Tin alloys that include lead must not also include cadmium, because together they form a low melting point eutectic, PbSnCd, melting at only 145°C.

Therefore, all alloys containing lead need in their specification a strict limitation of the cadmium impurity.

Until now this need has not been considered in the international standards specifications of white metals containing lead (e.g., the well-known WM80 or $SnSb_{12}Cu_6Pb$ in ISO 4381). The inadequate specification is one of the reasons for the alloy to sometimes be substandard, depending on the suppliers' qualifications.

Only the elements listed in the specification will be analyzed and their values documented.

Unspecified impurities remain unconsidered.

A decision to limit an impurity to 0.001% max. suggests at first sight a high-quality approach. But is this tight limit necessary from the technical point of view? Can it be achieved with sustainable effort? Remember that the tolerance of a test depends on several influences such as the accuracy of the reference samples, the tolerances of the test machine, the drift of the test machine, and the deviation of homogeneity of the tested material. If each of these individual tolerances is a power of ten smaller than the permissible impurity, then their combined effect may be 40% of the impurity limit. In this example alone, the homogeneity of the tested material has to be in a tolerance area of ±0.0001%. This clearly shows that the limit set is unrealistic, and for each repeat analysis the result will inevitably differ. The difference is greater still when the repeated tests are made on other equipment. The chances then are high that values will even be above the limit of 0.001%. In such a case, the specification is not met and there is endless discussion about the possibility of special release for the batch. That can all happen, although such a small impurity is insignificant for the quality of the material anyway.

For deciding the level of permissible impurity, the metallurgical need always has to be balanced against the practicable measurement technique.

10.2.2 Methods of Analysis

For manufacturing alloys with uniformly high quality, it is essential to use fast
and totally reliable methods of analysis. Speed is required so that the results
can be used for necessary adjustments during the alloying process and give a
consistent quality predominantly in the middle of the tolerance range.

The accuracy of the results for each element depends on the selected analysis
method. The highest accuracy is given by wet chemical analysis or Atomic
Absorption Spectrometry (AAS) and Inductive Coupled Plasma (ICP).
Within these methods, different qualities of measurement are attainable.
AAS is preferable for analyzing low concentrations; ICP is also good for low
concentrations (sequential modus of measurement) and for higher concentrations
(simultaneous modus of measurement). The best overall accuracy is given by
combining wet chemical with ICP, but this is slow and costly and therefore not
preferred for continuous quality control. The methods will typically be used for
determining the exact concentration of reference samples that are needed for the
production testing described next.

Methods of analysis requiring little preparation work and with measurement
results within a minute are the spectrographic analysis method of Radiograph
Fluorescence Analysis (RFA) or the Optical Emission Spectroscopic (OES)
procedure. Figure 10.1 shows a machine used for OES analysis. During the test
procedure, a spectrum is generated. With the RFA method, the surface of the
test material is activated by x-rays, and with the OES method, the spectrum is
generated by an electric arc. Both methods rely on comparison procedures. The
spectrum generated from the specimen is compared with the stored spectra of the
reference samples. Interim values are interpolated. For the interpolation to have
high accuracy, it is necessary for the test spectrum value to have two reference

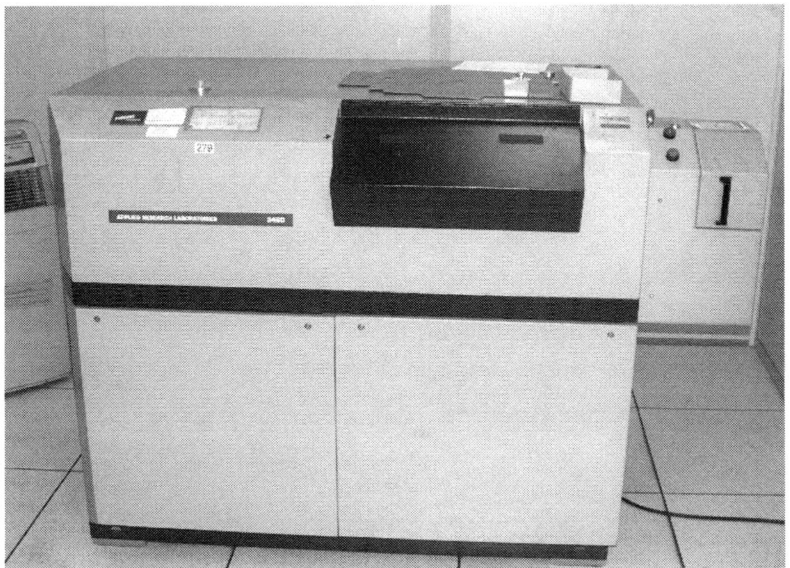

Fig. 10.1 Machine for OES analysis.

sample values as near as possible. The lack of linearity of the spectra means the quality of the result depends on the quantity and accuracy of reference samples. The greater the number of references, the better the result. Unfortunately, there are no certificated reference samples of white metal available on the market; they have to be produced by the end user in a complex exercise. In practice, just a few companies that are white metals specialists have sufficient reference samples to successfully use the RFA or OES procedure.

In the past it commonly happened that even accredited laboratories tested white metals and documented incorrect results on their analysis certificates. The reason was that they had only standards for the different elements such as tin, antimony, and copper but not for all possible alloying proportions and combinations. Consequently, their results varied up to 100% from reality. When a laboratory lacks sufficient reference samples to test a special alloy, then the spectral analysis must be excluded and wet chemical procedures preferred.

Finally, of course all test equipment has to be calibrated, and additionally, for RFA and OES, drift control is needed periodically.

OES and RFA analysis are comparison procedures.

OES and RFA analysis give usable results only when a multiplicity of reference samples is available within the range of measurement.

Decades of experience with both analysis procedures has led to the conclusion that for white metals the OES procedure gives the more reliable results. The electric arc penetration includes a bigger material volume and excludes local surface influences to the greatest possible extent. In particular, the unfavorable influences from surface preparation are reduced.

Certainly important for qualified test results is the homogeneity of the test specimen. For this reason, the specimen section as shown in Figs. 10.2 and 10.3 is used. The disc area used for the analysis is relatively thin and has a large surface; therefore, this area cools down very quickly in the mold and segregation is avoided. The geometry of the specimen excludes segregation influences from the upper cylindrical portion.

The determination of quantities of the different alloying elements happens simultaneously with RFA or OES analysis. Maximum accuracy is achieved by a minimum of three tests on each specimen. Averaging is made automatically, and the test with all results is complete after less than two minutes.

Test specimens for RFA and OES analysis must be homogeneous.

Frequently it happens that customers want to check the white metal against the supplied analysis certificate, and very often differing results are found. What is the reason?

The analysis certificate with the delivery is based on a specimen taken from the manufactured batch of liquid alloy. The specimen solidified homogeneously as

Fig. 10.2 *Test specimen positioned in the OES analysis machine.*

Fig. 10.3 *Lower surface of the test specimen. Two test sites are clearly visible.*

described. In contrast, the bulk white metal alloy has been cast into molds and there has solidified slowly, during which time segregation occurs in the ingot. This is unimportant because, when used, the ingots will be melted again and their analysis will in total be correct. When used for lining the plain bearing backings, the cooling and solidification is under control so that the final lining is homogeneous again.

The analysis made on ingots that are sure to be inhomogeneous cannot produce results conforming to the analysis of the batch. Great variations will be found when cutting an edge from the ingot and analyzing this. From experience, we know that at the edges of the ingots the segregation is worst and the concentration of the impurities greatest. Often there is also a high concentration of iron, which is abrasion residue from sawing.

OES or RFA analysis of white metal ingots is not representative and cannot fully confirm the analysis result of the batch.

Reasonably accurate results can be obtained by cutting the ingot across and spreading the points of analysis along the most trapezoidal section, as shown in Fig. 10.4. The average of these tests will at least provide a clear identification of the alloy type.

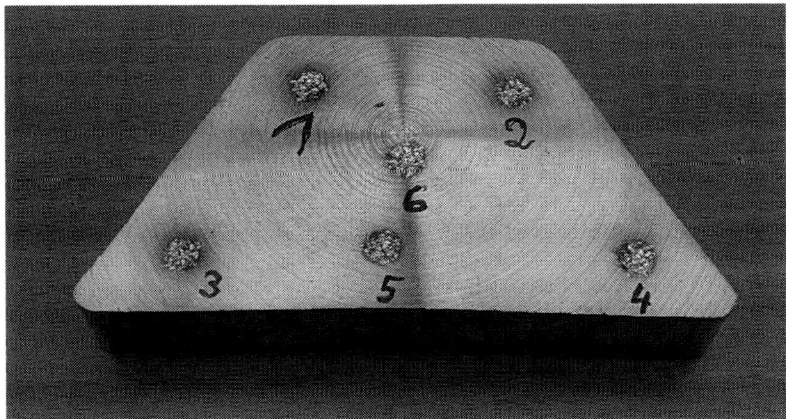

Fig. 10.4 Reasonable distribution of OES analysis along an ingot section.

From an ingot, an exact result is only possible by machining the ingot into small chips, mixing them thoroughly, and then making a chemical analysis.

If a single ingot is melted, and a small test sample taken for OES analysis, the small quantity means there is a risk of losing a portion of some elements by oxidation and compromising the result.

Often EDX analysis is selected. This supplies results of the existing elements at the tested spots, but it is not applicable for an average analysis of a white metal alloy.

Organizing an analysis is very complex. Selection of specimen, preparation, analysis procedure, and evaluation all need great care.

10.2.3 Recipe (Composition)

Consideration of the recipe, or alloy composition, is a further important quality feature not sufficiently respected by the industry. A baker knows very well what is needed for success with a cake. The correct quantities of different ingredients have to be used in the right sequence at the right temperature for the right cooking time based on the original recipe. Some components have to be vigorously stirred; others have to be slowly blended in. Only in this way will the end result be correct.

For industrial products the principle is the same, but there the knowledge about the recipe is called know-how. The assumption that the chemical analysis provides the correct recipe is wrong. The analysis certificate shows only the quantities of the included and tested elements. It gives no information about whether the sequence of adding the elements was correct, whether the tested material is an alloy or a mixture, and whether the elements are as crystals or in solution. For the correct quality, these significant points are controlled by the recipe or manufacturing procedure.

The analysis certificate confirms the quantities of the elements in the alloy.

The analysis certificate gives no information about the physical format of the elements in the alloy. That is controlled only by the recipe.

10.2.4 Reuse of White Metal Chips (Swarf)

During lined bearing manufacture, a high proportion of white metal is machined away as swarf. Economic and environmental considerations lead to a desire to reuse the material, and commonly white metal chips have been casually added to the new alloy in the melting pot. This action contains multiple quality risks and is not acceptable.

Segregation that occurs during spin casting means the upper layer of the lining is outside the tolerances, especially for copper and antimony, and this is the material that gets machined away. When reusing such material, the new casting will be completely out of specification.

White metal chips have a large, oxidized surface area. These oxides do not melt at the normal babbitting temperature of up to 540°C and therefore remain as impurities in the alloy.

The swarf will be contaminated with cutting oil, which produces intense and easily flammable gas. Residues from this will remain in the alloy and disrupt the normal formation of crystals. The crystals can grow to abnormal size and deviate from their natural shape.

CuSn crystals in particular will not completely dissolve when the alloy is remelted, even up to 540°C. Should such crystals get into the bearing, during machining they are crushed or completely broken out of the lining surface. Cracks occur—see Figs. 10.5 and 10.6. The local hardness of the running surface caused by these crystals is unacceptably high.

The cuboid SbSn crystals grow as a result of improper material recycling. A size of 15 times the original dimension is not unusual; see Figs. 10.7–11. Figure 10.12 shows for comparison a typical structure of pure new white metal.

Large malformed crystals tend to cause further problems by forming groups. Consequently, at the lining surface there are areas of crystal concentration with high hardness, and the remaining area is almost pure tin and accordingly soft. Under loading there is variable deformation, and the problem becomes clearly visible (see Fig. 10.13).

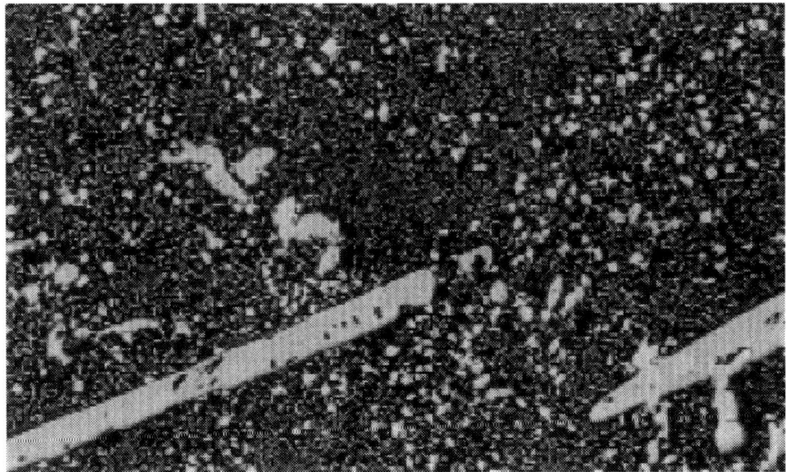

Fig. 10.5 *Extremely large Cu crystals. The other crystals are not homogeneously distributed.*

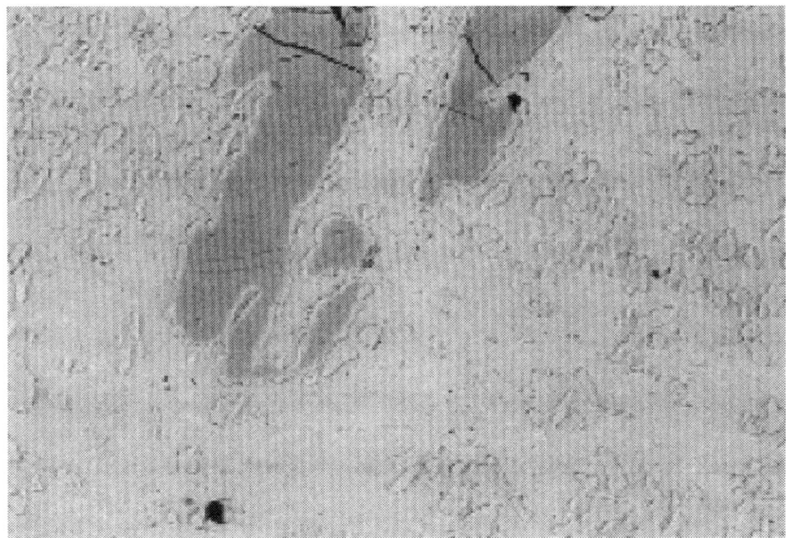

Fig. 10.6 *The oversize hard Cu crystals at the lining surface have been shattered by machining, and surface cracks result. The concentration of hardness at the surface is unacceptable.*

The influences mentioned as well as deviation of chemical composition lead especially to inhomogeneous distribution of crystals and hardness. Reduced load capacity and early crack formation result in a reduced service life of the plain bearing.

Even when melting white metal, swarf deviations may be noticeable. Due to unmelted crystals and oxides, the metal is too viscous. The crystals gravitate to the bond zone and reduce the bond strength between lining and backing. Working with the material becomes more difficult, and bond failures occur.

Fig. 10.7 *SbSn crystals that are usually cubic are here completely degenerated in shape and dimension due to incorrect recycling.*

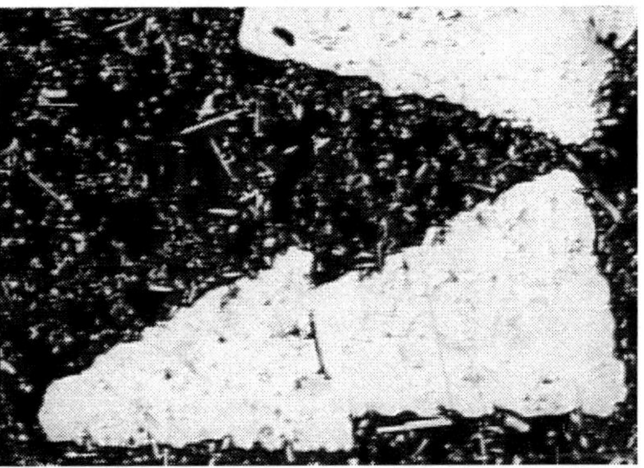

Fig. 10.8 *SbSn crystals that are usually cubic are here completely degenerated in shape and dimension due to incorrect recycling.*

Fig. 10.9 *The degenerated crystals also have a partly porous structure.*

Fig. 10.10 *Further examples of degenerated crystals due to incorrect recycling.*

Fig. 10.11 *Further examples of degenerated crystals due to incorrect recycling.*

Fig. 10.12 *For comparison, here is the structure of pure new white metal. The scale is the same for Figs. 10.5–10.12.*

Fig. 10.13 *Unevenly load-deformed surface due to variable hardness.*

Re-use of white metal chips introduces a significant quality risk. This risk cannot be nullified with the resources available in the plain bearing shop. Therefore it is strongly recommended to declare the chips as scrap (as prescribed by most legislators) and to dispose of it appropriately.

Adding of white metal chips to the plain bearing fabrication process is not acceptable.

10.3 Quality Control of Plain Bearings

To guarantee good and uniform quality of plain bearings, continuous observation of the fabrication process is needed. This includes quality control of the backings as well as the lining material. Most inspection effort concentrates on the finished bearing.

10.3.1 Inspection of the Geometry

Inspection of a plain bearing requires that all geometric data will be checked for conformity to the detail of the drawings and documented. Parts assembled for measuring must be marked so that after disassembling and later reassembling it can be guaranteed that the parts all occupy their original positions.

10.3.2 Inspection of the Bond

Besides checking dimensions and geometry, the quality inspection is always concerned with the bond between the white metal layer and the backing. There are destructive and nondestructive methods available.

The sensitivity of the bond strength to different marginal conditions has been clearly demonstrated. The question then arises: how does bonding work? Literature mentions that during the tinning procedure there is a chemical reaction, $FeSn_2$, resulting in the bond. This is accepted, but the explanation is not conclusive, because bronze backings also give good bond with tin, although, as is well known, no iron is present.

However, the fact is that there are many links between bond strength and backing material elements. With cast iron, it is easy to understand that depositions of perlite and graphite do not accept bonding. On steel backings, the elements chromium, nickel, manganese, and silicon have been discovered to have bond-strength reducing influences. It is likely that the bond quality also involves the concentration of these elements and their combination.

There is little knowledge about the interaction between bond strength and the alloying elements of the backing such as chromium, nickel, manganese, and silicon.

Often the compound of white metal and backing is explained as a diffusion process; sometimes it is called a soldering or welding process. Or is it a combination of all? Obviously it depends on the intensity and the temperature level, and at present we know very little about that.

With the conventional lining procedures, the compound is created in two steps. First, the backing surface is tinned; in this step diffusion obviously happens. Incidentally, this is only successful with pure tin and not when using white metal alloy. With the second step, casting the white metal, a melting process occurs between the tinning layer and the white metal. In comparison, when lining by laser, the tinning procedure is completely omitted and the bond is made directly between backing and lining material. The advantage of this technology is the controllable, and very consistently shallow, zone affected by heat. The bond strength produced by laser lining is much greater than with any of the conventional procedures.

To summarize: casting procedures can produce good bond strength, and with laser lining technology the bond strength is 50% better, with no tinning needed. The results are satisfying, but we do not yet know in detail what happens in the bonding process.

How bond strength develops under the different lining procedures is largely unknown. Nevertheless, in practice, good to excellent results are attainable.

10.3.2.1 Destructive Testing Procedures

There are many test procedures available (e.g., Chalmers, shear, bending, Erichsen, chisel, and twisting tests). The latter two procedures allow only a subjective examination of the bond quality and are therefore of limited value.

Destructive testing procedures can rarely be used for plain bearing quality inspection, but for R&D activities they are very important.

The Chalmers bond test is rated the best of the destructive methods because it quantifies the bond strength between backing and lining material. A specially fabricated specimen is ruptured at the bond intersection by a tensile force, shown in Fig. 10.14. The test can be made with white metal layers of 2 mm thickness and above. Considering the curve on a journal bearing, any specimen taken from the raw cast layer must be 1 mm thicker, and that means a minimum of 3 mm thickness. To accommodate the more intensive curve on smaller bearings, there are two different specimen geometries. The test procedure covers backings from steel- or copper-based materials, but not gray cast iron. The test specimens and procedure are defined in ISO 4386 part 2.

The destructive bond test defined in ISO 4386 part 2 no longer applies to backings of gray cast iron.

The author has tested hundreds of Chalmers specimens and found many inadequacies in the ISO documentation, beginning with the preparation of the specimen. The dimensioning and tolerances do not sufficiently consider the position of the bond intersection area.

In practice, this means that often the bond zone has a residual ring of backing material. When tested, this ring is torn from the backing, and the test result is invalid. The problem can be solved by machining the specimen (Fig. 10.15) according to Table 10.1 and following machining sequences from Table 10.2. The tolerances have been slightly modified from the standard, so the whole tested ring surface is closer to the nominal section value.

White metal layers less than 2 mm thick cannot be reliably tested by the Chalmers procedure. With thinner layers, the bending deformation starts to act in a cup spring effect. The rupture area is therefore subject to bending stress in addition to the intended tensile stress, and the results indicate reduced strength. ISO 4386 part 2 compensates for this influence by converting from relative bond to absolute bond strength. This approach goes in the right direction but should give greater consideration to the bending-stress influence.

The ISO standard does not sufficiently consider the position of the bond zone in the specimen. Often this leads to unusual results.

Fig. 10.14 Chalmers specimen for quantitative testing of bond strength between the lining and backing material. Left: before testing. Right: after the test.

With the standard test procedure, thin layers bend like a cup spring, and this adversely influences the test result.

The suggested specimen and test procedure modifications eliminate the problems described. The next version of ISO 4386 part 2 will consider this proposal (circa end of 2011).

With fabrication of the specimen according to Fig. 10.15 and Tables 10.1 and 10.2, the author achieved a clear definition of the specimens' geometry in the bond area and ensured that the top and bottom surfaces of the lining are parallel. These parallel surfaces are clamped in the modified test facility, as shown in Fig. 10.16. The additional bending of the lining is thus avoided, and the test results are accurate, even for thin layers. This procedure is already well proven with more stable results. When after conversion from relative to absolute bond strength the result exceeds the tensile strength of the lining material, the result has to be limited to the tensile strength. This logical restriction is not considered by the present standard.

The bond-strength test results vary because of the complicated specimen geometry and many other influences. Therefore, a minimum of three specimens should be taken from a bearing. It is helpful to take a specimen from the center as well as from the border area. The averaged result should never be understood as an absolute result, but as an indication of the level of the bond quality. It is therefore quite inappropriate to document bond strength to decimal places.

Absolute bond strength test results can be no more than the tensile strength of the lining material.

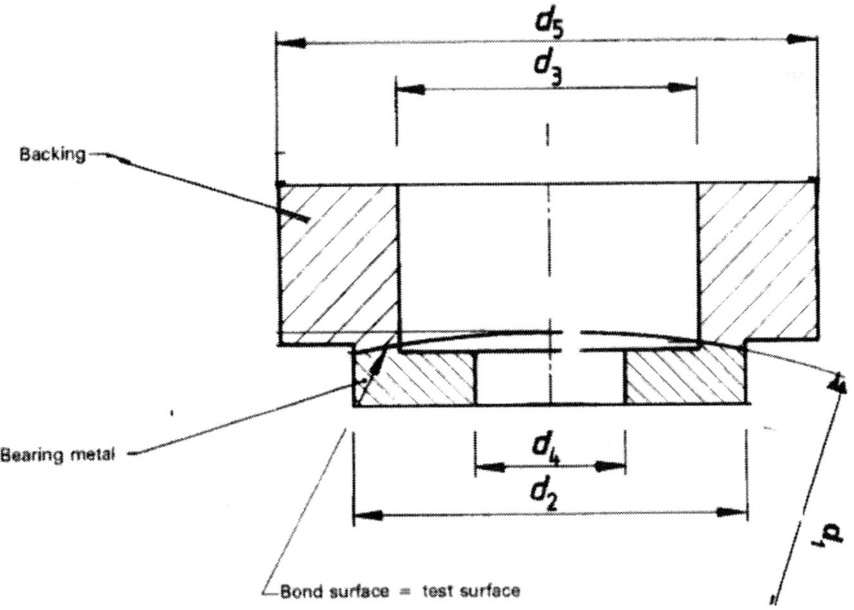

Fig. 10.15 *Chalmers specimen to ISO 4386 part 2.*

Table 10.1 *Proposed New Dimensions and Tolerances for Test Specimen*

Type of specimen	Bond surface	Bearing shell inside diameter	Specimen dimensions			
	A mm^2	d_1	d_2 h8	d_3 H8	d_4 +0.1 0.0	d_5
100	100	90 - 200	19.60	16	8.1	29
200	200	> 200	28.85	24	12.1	38

Table 10.2 *Specimen Machining Sequence*

	All data in mm	Specimen 100	Specimen 200
Step 1: a) tighten raw specimen with bearing metal toward chuck b) machine d5 c) machine front plane to thickness Z	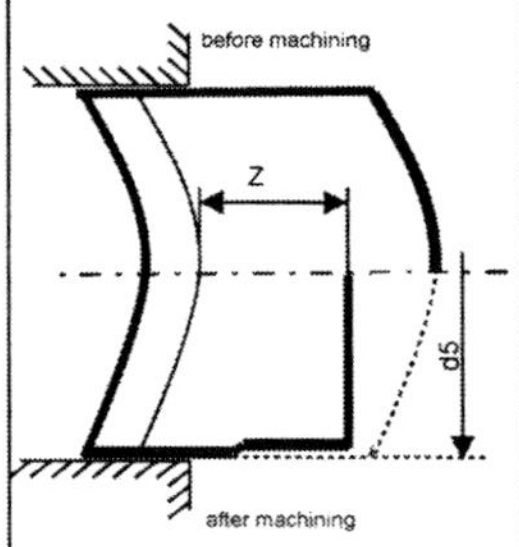	d5 = 29 Z = 8 - 10	d5 = 38 Z = 10 - 12
Step 2: a) reverse specimen and tighten with backing toward chuck b) drill d4 c) machine d5 machine d2 and machine the <u>reference plane</u> 0.1 mm deep into the backing referred to the deepest point of bond surface d) machine the bearing metal front face plane	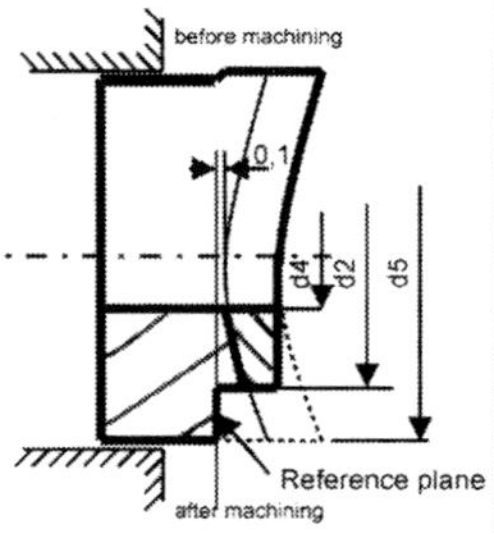	d4 = 8.1 $^{+\,0.1}_{\ \ 0}$ d5 = 29 d2 = 19.60 h8	d4 = 12.1 $^{+\,0.1}_{\ \ 0}$ d5 = 38 d2 = 28.85 h8
Step 3: a) reverse specimen and tighten with bearing metal toward chuck b) machine d3 with depth Z1	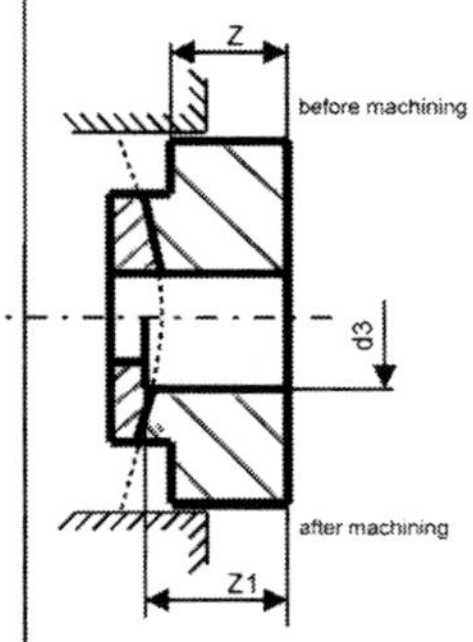	d3 = 16.0 H8 Z1 $^{+\,0.1}_{\ \ 0}$ = Z + 0.9	d3 = 24.0 H8 Z1 $^{+\,0.1}_{\ \ 0}$ = Z + 0.9

It is not very helpful to define by a standard just the geometry of the specimen and the test procedure without giving any guidance on assessing the test results.

When does the test result show good bond and when is the bond not sufficient?

Test results from the production line have to be differentiated from results of a qualification process. A qualification process should give better results so that, after the process optimization, the daily routine running fabrication should also show improved results despite some processing deviation.

It is not sufficient just to describe the specimen and test procedure. A guideline for evaluating the results is also needed.

Considering the described modified test procedure, the author recommends classifying the determined bond strength as good when the mean value of the absolute bond is at least 60% of the tensile strength of the lining material ($0.6 \times R_m$). For qualifying a process, the documented conditions and actions would be approved when the mean value of absolute bond is at least $0.8 \times R_m$.

The corrections presented here will be considered in the future in ISO 4386 part 2.

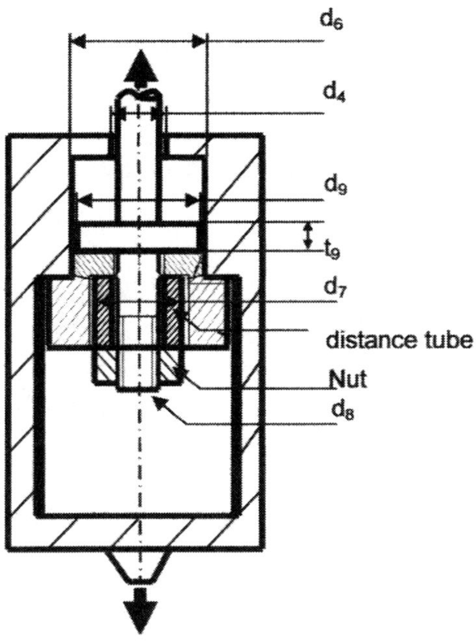

Type of specimen	The key dimensions of the modified test facility in mm					
	d_4 + 0.1 0	d_6 + 0.1 0	d_7 0 - 0.1	d_8	d_9	t_9
100	81	19.8	15.9	M 8	19	4
200	12.1	29.1	23.9	M 12	28	4

Fig. 10.16 *Modified test facility showing key dimensions.*

10.3.2.2 Nondestructive Testing Procedures

Because the destructive testing procedure—as the name says—destroys the bearing, the procedure is in only very limited use for continuous quality control.

There are some nondestructive testing procedures available, but unfortunately they all have limited informative value. The traditional ringing test allows at best a subjective estimation of large defects. To detect plane bond defects with x-ray absorption is not possible because there is no local decrease of material thickness. Electrically inductive procedures work with tin alloy lined on steel, but accuracy decreases with thicker linings, and on cast iron backings the procedure fails.

For quality inspection of plain bearings, there are worldwide only two nondestructive test procedures standardized and in regular use. The procedures are not perfect, but there are no better alternatives available. They are the ultrasonic testing procedure and the dye penetration testing procedure.

10.3.2.2.1 Nondestructive Ultrasonic Testing Procedure

In the production of plain bearings, the ultrasonic testing procedure is well established. It is based on the following physical rules:

At the boundary layer of two media, ultrasound is partially reflected. The bond interface between lining material and backing, and the transition between backing and atmosphere, are such boundary layers that produce measurable reflections. From these it is possible to detect the existence of bond for each backing and lining material combination independently of the thickness. Various ultrasonic procedures are in use.

The sound transmission procedure is applied for thin-walled plain bearings or for backings with high sound attenuation. The system uses two probes mounted one each side of the test area. One probe continuously transmits sound into the plain bearing, and the opposite probe measures received sound intensity. In any area of bond defect, the sound transmission is interrupted. The interruption cannot, however, be located precisely in the wall thickness, so it may be uncertain whether the indication comes from defective bond or some flaw or feature in the backing such as an oil-way.

The pulse-echo method has been standardized for plain bearings by ISO 4386 Part 1. The method uses normal beam probes working simultaneously as transmitter and receiver by sending an ultrasonic pulse into the tested material and measuring the time delay of the echo from the bond zone or back-wall surface. The sonic speed in different materials is known, and this with the echo time permits the distance to be calculated. In practice the screen display receives the calculated data, so the distance of the detected bond defect from surface or from back-wall can be shown exactly.

This procedure does not detect bond defects less than half the diameter of the ultrasonic probe because of undefined reflections. The test can be performed on metallic multilayer plain bearings consisting of steel- or copper-based materials lined with tin- or lead-based bearing metals. Selection of an appropriate probe will also enable bond defects in thin layers to be detected.

The test surface must be clean, and the surface roughness should be $R_a \leq 5\ \mu m$. For good contact between ultrasonic probe and the bearing surface, a wetting with light machine oil is recommended. Small bearings and bearings with extremely thin linings are best tested by immersion scanning in a water bath for consistent probe contact.

It is essential for valid results that the contact surface, bond interface, and the back-wall surface be parallel, because inclined surfaces prevent ultrasonic reflection back to the probe. This effect also happens along the edges of dovetail grooves, making detection of bond defects there impossible. This is unfortunate, because there is a high probability of defects in this area. Such geometrically critical areas, as well as porous surfaces on the backing, should be documented before lining to avoid later misinterpretation as bond defects, although genuine bond defects cannot be ruled out at exactly these positions. These problems are particularly severe with gray cast iron backings, for which we have the most difficulty achieving good bond anyway. It is, therefore, clearly not very helpful to use ultrasonic bond testing on gray cast iron backings. Steel backings with ungrooved plain bond surfaces make good bonding practicable, and the whole area can be accurately tested. This configuration therefore gives the maximum possibility of good quality.

Dovetail grooves and surface porosity in gray cast iron backings must be documented before lining. This way, later ultrasonic indication at these positions can be anticipated. However, bond defects at the same position cannot be fully excluded.

Ultrasonic testing of bond on gray cast iron plain bearings does not give conclusive results.

Usually the preferred direction of testing is from the bearing metal side. It is recommended to machine the cast lining layer to get an even surface for testing. Small bushings with inside diameter < 50 mm have to be tested in the opposite direction, from outside the backing.

The ultrasonic instrument should be fitted with a calibrated attenuator, reading in decibels, and have adjustable time-base ranges.

For standard procedure, normal beam probes should be chosen with size and frequency according to the size of the bearing and requested test accuracy, thickness of the lining, backing thickness, and backing material. The typical probe size range is from 10 to 24 mm diameter, and the frequency is about 4 MHz. For lining thickness above 10 mm, and backings of gray cast iron or lead-bronze, 2 MHz is preferred because of better sound penetration. For thin-walled bearings, 12-MHz probes are also in use.

For most normally loaded bearings, a complete scan of the running surface is sufficient. For high-duty bearings, the complete scan should include all contact areas such as running surface and flanges. The highest loaded areas should be scanned with 20% overlap to ensure that the complete bond area is covered. The ISO 4386-1 defines different test categories.

Before starting the tests, the time base range and the sensitivity should be adjusted on the screen according to the thickness of the tested bearing. Usually the test is made with a back-wall echo. When the back-wall profile is unsuitable, for example a spherical seating, the test can still be done but the procedure is different. Unfortunately, in the old ISO standard the explanation of both procedures is wrong and misleading, and it will be updated soon, as described here. Here follows the definition of both procedures:

Testing with back-wall echo:

- Preconditions for the test with back-wall echo are:

- Free access of the probe to the white metal layer

- Back wall is always parallel to the lining surface

- No hollows or voids inside the backing material

The time base range and sensitivity shall be adjusted so that at least the first back-wall echo is visible on the right side of the screen at approximately 80% of screen height. If in doubt about having the first back-wall echo on the screen, use a reference block for calibration. The bond echo is on the left side of the screen near the input signal.

Defective bond is indicated when the back-wall echo breaks down and at the same time the bond echo at the left of the screen increases and repeats itself; see Fig. 10.17. The border of the detected defect area is in the middle of the probe diameter when the back-wall echo is reduced to half its height under the previously mentioned conditions.

Testing without a back-wall echo:

If the preconditions cannot be met, testing without a back-wall echo becomes necessary. For this procedure, use a reference block with bond between similar backing material and similar bearing metal lining as for the tested bearing.

Using the reference block, the bond echo should be set at approximately 20% screen height; see Fig. 10.18. Bond is shown on the bearing when the bond echo is at the 20% level shown in Fig. 10.19. A bond defect is present when the bond echo from the bearing is significantly more than 20% screen height (see Fig. 10.20).

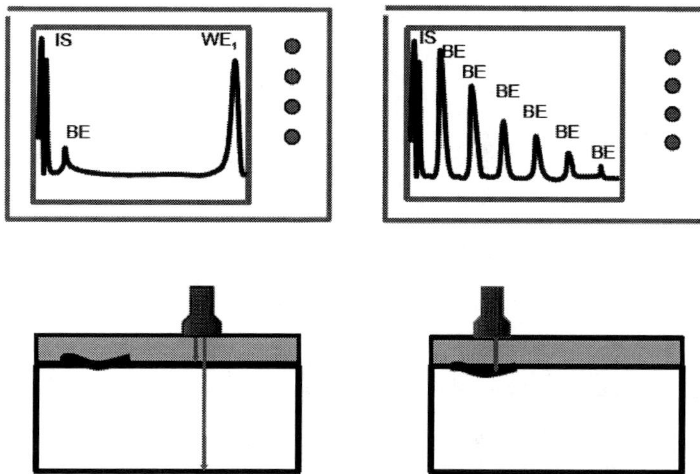

Fig. 10.17 *Testing with a back-wall echo. On the left with bond: small bond echo and big back-wall echo. On the right with bond defect: big and repeated bond echo, no back-wall echo.*

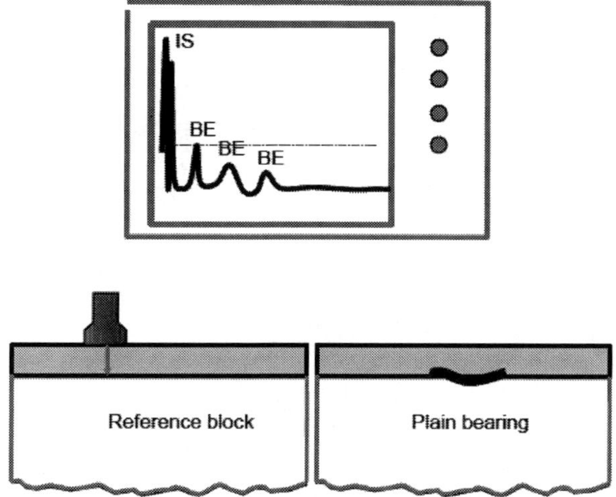

Fig. 10.18 *Testing without a back-wall echo: adjustment of the bond echo to 20% screen height using a reference block.*

Generally, when evaluating the bond, all defects are marked by straight boundary lines. The location of the center of the probe is decisive for determining the transition line between bond and no bond. Isolated point-type defects should be marked with a value equal to half the probe diameter. If the distance between two or more defects is less than 10% of the bearing width, then these defects are regarded as one continuous defect.

Table 1 of ISO 4386-1 lists various defect groups. These recommendations are problematic. First, there is the faulty description of defect dimensions along the

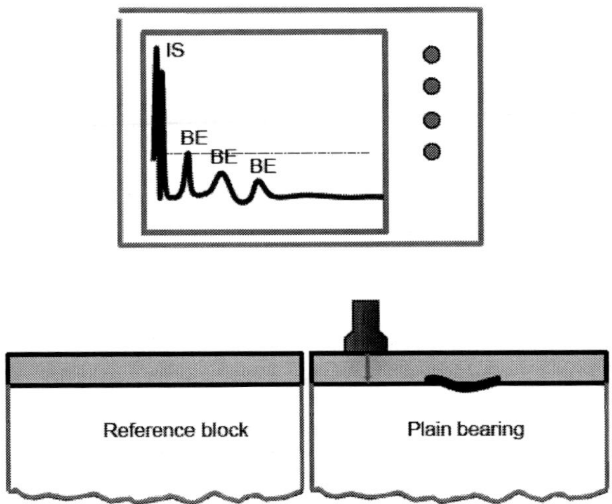

Fig. 10.19 *Testing without a back-wall echo: display of bond similar to the result from the reference block.*

bearing edge zones. That is the visible bond line of the backing to the bearing metal on end faces or joint faces. This zone is less than half the probe diameter, so it cannot be tested by ultrasonic procedure at all, and such defects have to be tested by the penetrant testing method specified in ISO 4386-3.

Defects along bearing edge zones as defined on ISO 4386-1 cannot be found with the ultrasonic test procedure at all.

The practicality of the definition of single and total defects in ISO 4386-1 should be reviewed. The problem is that with increasing bearing size the maximum permissible single defect increases linearly because it is related to the bearing width. In contrast, the permissible total defect increases disproportionately because it depends on the bearing surface. So with increasing bearing size the permissible number of single defects increases rapidly. For example, here is a comparison of two bearing shells based on defect group B1:

A small shell with inside diameter 100 mm and width 60 mm has a running surface of 9.420 mm². For this shell, the defect group B1 allows a permissible total defect of 94 mm² and a maximum single defect of 45 mm². That is a single defect of nearly 7×7 mm. This maximum single defect is permitted twice per bearing.

A bigger shell with inside diameter 600 mm and width 360 mm has a running surface of 339.000 mm². Here, the permissible total defect is 3390 mm², and the maximum dimension for a single defect is 270 mm². That is a single defect of nearly 16 × 16 mm, and this maximum single defect is allowed 12 times per bearing. If the single defects are smaller, for example 10 × 10 mm = 100 mm², then 33 single defects are permitted on the shell by defect group B1. For defect group C, as many as 67 defects of 10 × 10 mm or 9 defects of the maximum size of

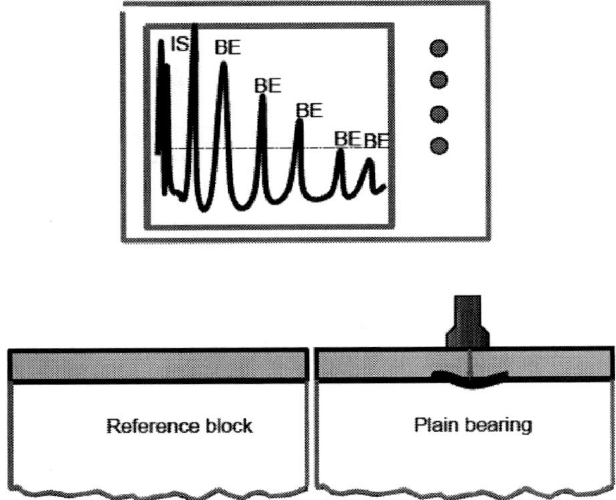

Fig. 10.20 *Testing without a back-wall echo: when the bond echo is significantly bigger than the reference (on 20% screen height), then there is a bond defect.*

27 × 27 mm are accepted. For defect group D, 11 single defects of 38 × 38 mm or 170 defects of 10 × 10 mm are permissible.

It must not be forgotten that here we refer to permissible defects which need not be corrected! Furthermore, it has to be considered that the ultrasonic test procedure can only show missing bond. The test procedure gives no information whether the bonded area has high or low bond strength. From practice, it is well known that multiple bond defects in a bearing give the clear hint that the bond quality is generally on a low level. The occurrence of defects such as from the example of the big shell should not be defined as permissible, but should be taken as an indication of weak bond strength. With the present definition of bond defects in the ISO standard, it is not surprising that customers prefer to specify defect group "A" that allows no single defect. That is the other extreme, and it is neither technically reasonable nor economic.

It is urgent to adapt the definition of the defect groups, perhaps such that the single defect area may grow along with the dimensions of the bearing, but the permissible number of single defects should be limited according to the selected quality group. Then a definition of total defect area is no longer needed. Unfortunately, there is no support from the plain bearing industry for this or any similar adaptation.

The ISO standard also gives no guideline for interpretation of defects found and how to deal with them. For example, concentrations of many defects are clear hints of low bond strength in an area. In contrast, single defects on end faces or joint faces can be repaired in most cases by qualified soldering procedures under the proviso that a repeat test will confirm bond after the repair work.

The definition of defect groups in DIN ISO 4386 part 1 allows a risky quantity of defects. The corresponding inference of low bond strength is completely missing.

The test procedure should be carried out by a qualified person with not just a good education in ultrasonic testing but an additional qualification in testing lined plain bearings.

The author has met many excellent UltraSound operators specialized in testing welding or steel parts but who failed with testing of compound plain bearings because they have never been trained for them. Sometimes the soft lining layer has not been considered, so that the surface became badly scratched by the probe, and by the end of testing the bearing was destroyed. If the operator is not educated for testing compound bearings and the relevant standard is faulty, then the situation becomes problematic.

Operators for ultrasonic testing should be qualified in compound bearing testing.

Investigations have been made and concluded that the displayed echo height is in direct proportion to the bond strength. Theoretically that might be so, but in practical use the procedure shows clearly that the height of indicated echo continuously varies. That is due to the unavoidable variations in coupling between probe and plain bearing surface. Therefore, in reality interpretation of indications cannot quantify the strength of the bond.

Presently the standard ISO 4386 part 1 is under revision with the aim of integrating the amendments outlined here.

The unequal coupling between probe and bearing allows no further interpretations beyond the described detection of bond defects.

The ultrasonic test procedure is the preferred test method for detecting bond defects simply due to lack of alternatives. The procedure does not supply any quantitative information of bond quality.

It is desirable to develop a nondestructive test procedure for quantitative evaluation of bond strength.

10.3.2.2.2 The HOYT® Bondmeter (5)

The principle of operation is centered on a precise micro power source that causes current to flow through the work being examined. In the case of a bond defect, the current flows in the white metal only, resulting in higher electrical resistance, which changes the LED indication from green to red.

The unit self-calibrates when the contacts are pressed on an area without bond defect. When the LED shows green, the operator presses the on/off button, a beep sounds, and the calibration is concluded. The HOYT® Bondmeter (5) is compact and simple to use with the red/green indications for bond defects.

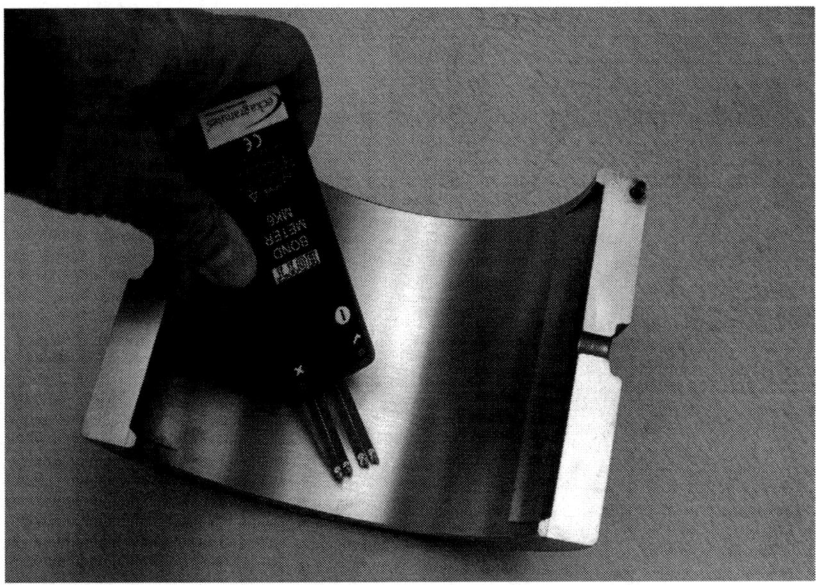

Fig. 10.21 HOYT® Bondmeter (5).

Because the HOYT® Bondmeter (5) does not use ultrasonics, it cannot be accepted for final inspection based on DIN ISO 4386 part 1. The relatively wide probe head restricts the exact location of very small defects, and so this machine is not a replacement for the ultrasonic testing method, but it is ideal for intermediate testing during bearing fabrication and needs no specialized training.

10.3.2.2.3 Nondestructive Dye Penetrant Testing

Dye penetration processes are used for nondestructive detection of surface defects such as pores, cracks, bond defects at edges, and similar defects. For this procedure to work, a defect must be on the surface or in direct connection with the surface. There is no limitation on the materials, but they must not normally be porous and should not chemically react with the test medium.

Quality control on plain bearings should be based on ISO 4386. The penetrant testing procedure is included there in part 3. It should be used for detecting bond defects on the transition areas between backing and lining layer, such as on joint faces or flange sides. These areas cannot be tested by the ultrasonic procedure.

Because the penetrant testing procedure allows checking of areas that are not suitable for ultrasonic testing, the two procedures complement each other.

Dye penetration testing should be carried out after final machining of the bearing.

The process is as follows:

Liquids penetrate into porosity open to the surface, cracks or bond defects, by capillary action. After a time, the liquid is partially sweated out.

All dye penetration test methods use this effect and emphasize it by using absorptive developers. After penetration by the test liquid, the developer is applied to the surface to clearly show concentrations of the test liquid in defects.

The surface to be tested has to be thoroughly cleaned of dirt, grease, and oil to allow the unhindered penetration of the test liquid. Just to clean mechanically by brushing or blasting is not sufficient. The best method is to clean the test piece in a solvent or ultrasonic cleaning bath. The cleaning agent used must be compatible with the test liquid so that the washing procedure does not affect the penetrant. The item is then dried, and the red test liquid (penetrant) is applied. Smaller parts can be dipped completely into the test liquid. For big parts, just the areas in question are painted or sprayed with the test liquid. After sufficient penetration time—usually 10 to 20 minutes—the surplus liquid is removed from the surface completely. In the next step, the developer is sprayed on the surface. It is a white and quick-drying medium with good absorbent behavior. The test liquid in the defects is drawn out and visible as red indications on the white developer. The defect areas show an intensive bleeding effect over time, which means the exact dimension of a defect cannot be determined by this method, and a long developing time often leads to misinterpretations when evaluating. Thus, the evaluation should be made within thirty minutes of spraying the developer. Later evaluation always leads to wrong interpretation.

The procedure also is useful to find pores and cracks on the running surface of bearings in service. For new bearings, it is not recommended to test the complete surface because pores are visible without the test procedure, and the running surface on new bearings is usually free of micro cracks.

Defect evaluation in the dye penetration test has to be done not longer than thirty minutes from spraying with developer. Later evaluation will lead to wrong interpretation.

10.3.3 Customers' Specifications

Many customers have their own specifications for the plain bearings they order, and specific acceptance procedures have to be fulfilled by the supplier. This is done to get the best possible quality, but often the request is excessive. The outcome may not be as intended. Sometimes the request results in unnecessarily high workloads, which make fabrication expensive and uneconomic. In the long run, it is always the customer who meets the cost of excessive requests!

It is a false assumption that quality requests always result in better quality. For example, when a customer requests centrifugal casting for a bearing with over 200 mm backing thickness, and refuses the solder lining procedure as an alternative, then he takes the risk of getting a worse result.

Quality tests should be based on relevant standards, but sometimes the standards insufficiencies need to be compensated by the experience of the plain bearing producer.

With the incoming goods acceptance inspection, the customer usually repeats all the documented dimension measurements made by the supplier. When the measured data conforms to the drawings, it is usually assumed that everything is in best order.

That may not always be true if checking the actual dimensions of the assembled bearing unit is neglected. Particularly in the case of repair work, checking of clearance and concentricity of the housing bore should not be forgotten.

When bearing damages happen immediately after assembly, it is common to immediately assume that the fault lies with the bearing supplier.

Practical experience, however, has taught us that these damages are time and again based on deviations on site or on assembling faults. Therefore the customer is well advised to direct his attention to these influences.

Functional efficiency of a plain bearing depends on the geometry of the bearing as well as on the geometry of the complete assembly. Therefore, it is not sufficient to limit the quality inspections only to the plain bearing.

The importance of considering the original white metal production formula has already been pointed out. As mentioned in Section 10.2.3, on one hand the supplier will not publish the recipe, and on the other hand the analysis certificate gives no information as to whether the original recipe was considered. Plain bearing producers and consumers should use alloys from suppliers that they trust and have found reliable.

This applies not just to white metals but also to other materials. Such problems are also well known from steel. Although material name and certificates are in order, sometimes there is a surprise when using the material. With changing or unknown origin of material, strange problems can be intermittent and impossible to trace back.

For special alloys it is doubly important to use the original, because only the developer of the speciality has the original recipe. By using original products and original technologies, the bearing producer obtains many advantages. He is in the position to produce high-quality bearings. Using the original manufacturer's innovative ideas, customers are sure to enjoy a competitive edge. Users of original technology show their professionalism. They also benefit from the manufacturer's depth of experience. When using fakes, problems arise with patent infringement and other issues. Often even the brand names are misused, and the supplied product is absolutely not in conformity with the genuine original. When using such a product in good faith, big problems of product liability occur between customer and bearing supplier.

To be sure to obtain original products, certificates of the materials used should always be requested to accompany the delivered bearing. Direct contact with the white metal supplier can provide easy confirmation of whether the supplied certificate is an original one. Large consumers who buy their bearings from the international market already successfully use this additional control to ensure they always get what they order.

Customers are well advised to request, together with the supplied plain bearing, the corresponding certificate of the lining material.

The producer of the original product can check the originality of the analysis certificate.

10.3.4 The Quality of the Standards

The relevant standards for testing plain bearings have been discussed together with some comments on inadequacies. Recommendations for improvement have been given. A similar situation occurs in Chapter 12 with the plain bearing damages.

The question arises of why standards that are permanently in use have mistakes and why the mistakes are not removed.

The phenomenon depends on:

- the qualification processing of a standard

- the age of an existing standard

- the structure and the regulations of the standardization organization

The quality of a standard obviously depends on qualified work. That does not mean that a defective standard is the result of unqualified processors. Actually the reason is that the processors often do not get the needed expert support. In practice, a small group tends to produce all the standards of a specified field (for example, plain bearing technology), and they often do not receive the needed expert support from the industry. There is the tendency that the industry on the one hand uses the standards, but on the other hand they distance themselves from investing money and manpower in producing and improving standards.

It would be very helpful to receive temporary expert support from the industry for solving the problems arising. Unfortunately even that is missing.

A big problem is the financing of the standardization work. The industry already has costs when they send employees for standardization work. Additionally, they have to finance DIN or ISO institutions with membership fees, and the rules are difficult in that the fees rise proportionally to involvement. When an active member has to buy a standard before being able to vote for a five-year revision, then it is not surprising that the number of members goes downward.

Standards should be primarily financed by the users and not by the producers.

If employees are available, it is relatively easy to develop a new DIN or ISO standard. Much more difficult is the revision of an existing standard based on the system of the five-year review. For an existing standard, it is assumed that it is proven, and the voting to continue the standard usually goes ahead without detailed checking. If an individual checks in detail, finds mistakes, and votes for revision, then he gets voted down and the standard remains unchanged. The problem is the voting system, with voting first and discussing later. On this basis, revisions become endlessly delayed. Although a member of DIN advisory committees and some working groups, the author had no chance to influence in a positive way the efficiency of standardization work. For some that leads to resignation. Finally, the industry has it in hand to exert more influence by intensive involvement in the standardization work and the results.

Standards are exactly as good or as bad as the level of industrial involvement.

Chapter 11

Plain Bearing Assembly and Operation

Essentials for long bearing service life are correct assembly and observation during operation. When assembling, the following demands have to be considered:

- No distortion

- Full contact between bearing and housing

- No damage and no contamination

- No misalignment

- Full lubricant supply

There is often a smile at the hint of the above demands, because all the items are understood. However, practical experience is that always when an assembly problem results in damage, one of the mentioned items was not considered. When this happens nobody smiles.

During assembly the highest accuracy is required.

11.1 Assembling Without Distortion

The need for unrestricted assembly is clear. Physical restrictions can distort the geometry of the bearing and endanger the operation.

Distortion can occur in different ways. Figures 11.1 and 11.2 show the typical problems diagrammatically. In practice, these influences are often, unfortunately, not so clear. The simplest case is the poor fitting or incorrect clearance between bearing and housing; see Fig. 11.1.

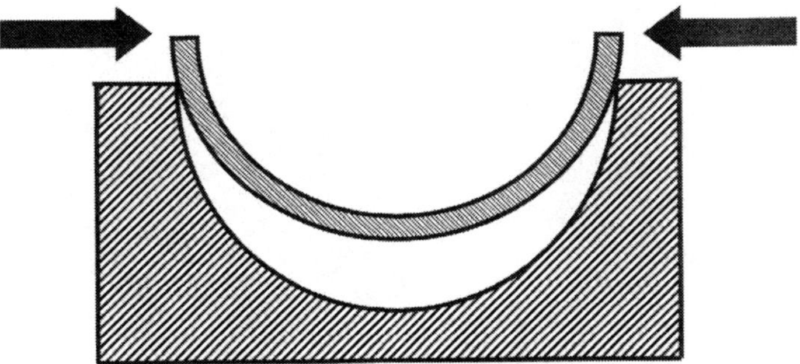

Fig. 11.1 *Distortion occurs when geometry of bearing and housing do not match.*

Bearing shells distort when the bolts joining the two halves have not been uniformly tightened or when the joint faces are not parallel. Bearing shells should be fixed and centered to avoid offset or twist of the halves during assembly (see Fig. 11.2).

Both shells of a split bearing require clear marking to identify the original orientation. If there is any doubt, the bearing should not be assembled. The same applies to a split housing.

11.2 Seating of the Bearing

A plain bearing is, together with its seating and housing, a unit. Therefore, it is not sufficient to check just the dimensions of the bearing, but the housing bore must properly fit as well. Especially when replacing bearings during repairs, the precaution of checking whether the bearing is concentric with the seating avoids some unpleasant surprises.

For example, on piston compressors, deformation of the housing bore often stems from previous damages. Seated in such a deformed housing, even the best-produced bearing has no chance of survival.

To avoid distortion, the geometry of the bearing and the corresponding housing bore must be checked systematically.

An oval bore cannot support the bearing uniformly and, in the direction of loading, there are zones with no contact, so the bearing is subject to additional local bending deformation. This leads to early fatigue, starting with cracks on the white metal surface. Such voids between bearing and housing bore can also be indicated by frictional corrosion on the bearing outside diameter.

The assumption of correct conditions of the installation may be wrong and may lead to a failure.

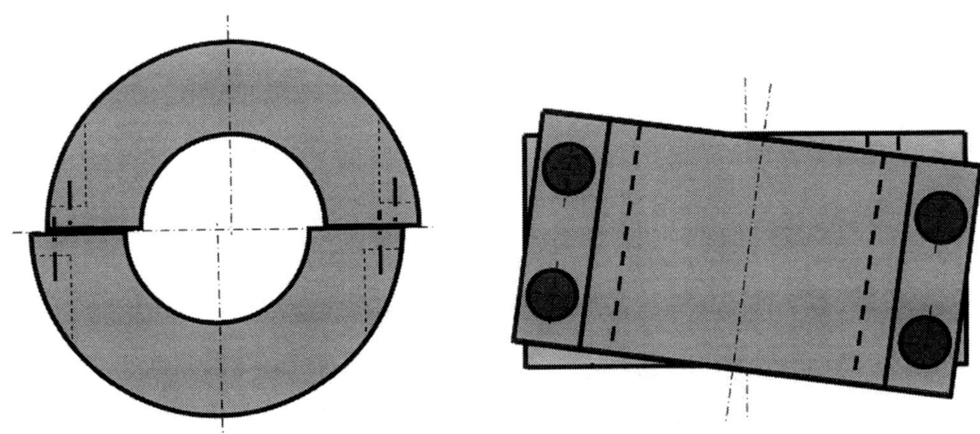

Fig. 11.2 Offset or twist of the bearing halves will lead to damage.

11.3 Damages and Contaminations

It is clearly unwise to use bearings damaged during transport or assembly. Scratches or dents should be professionally removed. If there is any doubt, use a replacement bearing. The most dangerous supply damages arise during assembly as a result of carelessness (e.g., contamination by flying sparks from nearby grinding work or damage when a tool falls onto the bearing). On site, the assembly area and the bearing itself must be carefully protected against damage and contamination.

During assembly work, the risk of damaging the bearing is particularly high. Appropriate actions have to be taken to exclude the risks.

Plain bearings must be free from contamination, which is easier said than done. Contamination problems are always based on insufficient care and may happen before and during assembling.

Contamination occurring before assembly may be shot-blast material, particles of paint, or machine swarf. Unfortunately, such contaminations are not usually at the surface but are hidden in grooves or in the lubrication system passages.

Contamination during assembly may be from grinding work or any kind of airborne dirt from the surrounding area—and, of course, the infamous cleaning cloth may be found casually discarded in the oil circulation system.

Interruption of assembly work or a crew change during assembly may lead to fatal damages. One worker is reworking an oil grove. During his working break the next worker, not informed about the activities, fits the rotor, and the residual swarf damages the bearing; see Fig. 11.3. Afterward, everybody acts smart and cannot understand how such an event may happen. The answer is always the same: insufficient care.

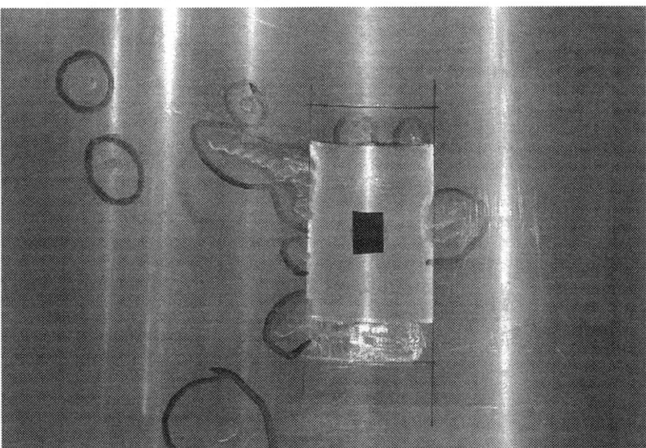

Fig. 11.3 Rework on an oil grove on site. The machinings have not been removed before the rotor has been installed. The marks are clearly visible on the running surface.

11.4 Misalignment

To check the alignment between a single bearing and the shaft is still relatively easy. Considerably more difficulty is experienced in the situation when a shaft is supported by several bearings. On big machines, the elastic curve of the shaft has to be considered, possibly additionally influenced by a coupling within the system. In such a case, the divergent bending under operational conditions has to be considered as well. Problems are most likely to arise when real conditions differ from the theoretical, and only the theoretical conditions are being evaluated.

Here is a typical example:

On a multi-bearing rotor system consisting of turbine and generator rotors, the bearings were replaced at an overhaul after many years' service due to intensive wear. The new bearings were accurately assembled, and all built-in parts were checked. The bearing supports and their positions relative to each other were not checked because it was assumed that their position remained correct as documented on the drawings. However, over the years the foundations had settled and the bearings had worn selectively, thus adapting to the new alignment, and so the foundation settlement remained undetected. The new bearings were fitted in the old housings and were off center by about 1 mm due to the settlement. The result was an immediate breakdown when the machine was started.

11.5 The Lubrication Supply

Proper lubrication is essential for safe bearing operation and is often the subject of the first questions in cases of bearing damage. The ever monotonous answer is, "There was sufficient oil pressure." This answer does not, however, automatically confirm a sufficient oil supply quantity. Think about a hydrostatic system blocked at the entry point into the bearing: pressure is there, but no oil supply!

Measured oil pressure is not a guarantee of proper oil supply.

The oil supply therefore deserves full attention. During bearing assembly, the oil feed system must be checked to see that it is completely filled and works. A hydrostatic system has to be tested under standing conditions before the first machine start. Use a test gauge for checking for shaft lifting. With this test, several possible problems are simultaneously excluded:

- Any possible obstruction in the oil feed system will be detected.

- After the test, the system is vented and filled completely with oil so on the first start the bearing gets oil from the beginning.

On bearings with hydrostatic jacking for start up, whether the non-return valve works properly has to be checked. If not, the hydrodynamic pressure declines after switching off the hydrostatic pressure, and the lubrication breaks down completely.

Finally, it must be checked whether the machine has been filled with a sufficient quantity of the correct oil grade.

11.6 The Scraping Procedure

Scraping is a cutting procedure used to achieve a very even surface almost clear of any tool marks. Usually scraping is done manually for the finest surface with high uniformity. Using the tool with systematic changes of direction, a typical structure will become visible on the surface (see Fig. 11.4). This pattern embellishes the surface and makes later wear clearly visible. To achieve good results using a scraper requires skill, experience, and especially plenty of time.

For improving the evenness of a surface, a reference plate, or for cylindrical surfaces a mandrel gauge, will be needed to make raised areas visible for scraping.

Fig. 11.4 *Pad of a thrust bearing with a hand-scraped surface.*

Today's plain bearings are supplied ready for installation. They have the correct dimensions and surface quality, and there is no need for additional scraping to improve the quality. Despite this, there are a few service technicians who staunchly defend the old tradition of hand scraping to produce artistic surface patterns. There is no technical advantage with this time-consuming procedure. Quite the contrary, it is possible by surface scraping to remove sufficient quantities of material to adversely affect the geometry of the bearing. In the end, the functionality of the bearing is reduced.

Surface pattern scraping is no practical improvement for plain bearings.

Here are two examples:

For a plain bearing system, bore and shaft have to be compatible. This requirement is fulfilled by the machining, and the tolerances are maintained with high precision. If the configuration is changed during assembly by some distortion, then at best the bearing can be adapted to the shaft by scraping. This does not solve the problem, as the bearing is no longer circular, and the adaptation has affected the lubrication gap geometry. The reason for the failure is the lack of a reference for the bore during the scraping procedure. By using the shaft for reference, there is a line contact because of different diameters of shaft and bore. The contact area can be varied by moving the shaft in the bore. By scraping in this area, finally an adaptation of the shaft to the bore happens, similar to the damage caused by mixed or boundary lubrication. The geometric conditions for proper hydrodynamic operation become systematically destroyed by this action.

Scraping of surfaces without real references, such as a surface plate or mandrel gauge, gives an inferior result. The shaft is no mandrel gauge.

In practice, bearing and housing sometimes suffer adverse deformation. The reason is lack of rigidity at the bearing supports or unexpected extremes of applied load—see Fig. 11.5. (Figure 11.6 shows damage caused by distortion.) This is another problem that cannot be solved by scraping. Another cause of deformation is that during assembly the bearing experiences only its own weight. When operating, there are completely different load and deformation conditions, and in designing the machine, the range of deformation has to be considered. For example, on large cement mill bearings the problem is addressed by designing a locally increased clearance, as shown in Fig. 11.7. With this design, the sector provided with hydrodynamic lubrication gap geometry has been minimized to the needed area. Outside this functional area, the gap has been increased to avoid pinching due to deformation.

Operational reliability of a plain bearing has to be calculated and designed for all operating conditions. This is the precondition for ready-to-install bearings. Scraping on site is no alternative.

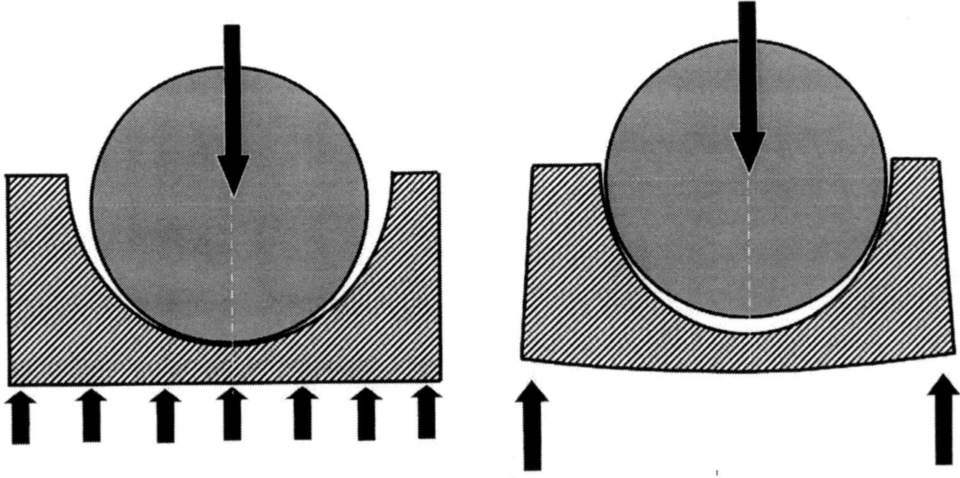

Fig. 11.5 *On machines with operating load, bearings and supporting structures may suffer deformation. This can affect the functionality of the bearing.*

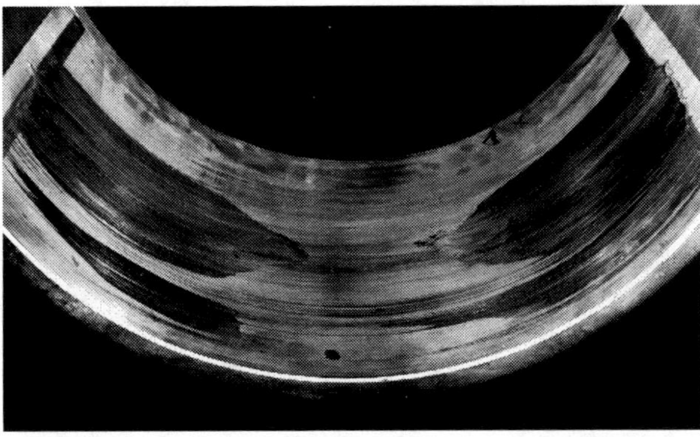

Fig. 11.6 *Typical damage due to distortion: clearly visible is the wear caused by boundary lubrication near the joint face area. The bearing worked like a "drum brake."*

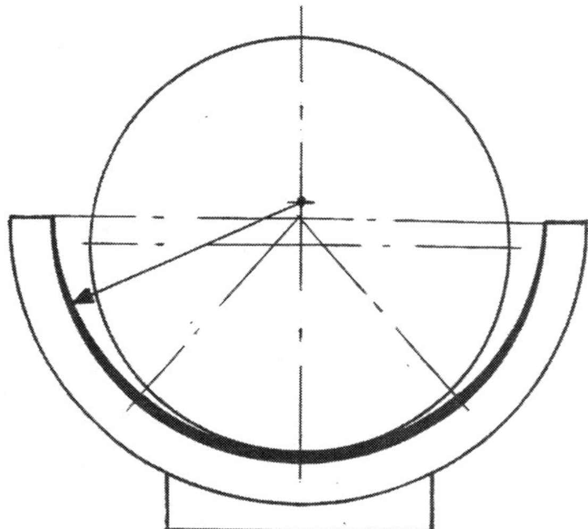

Fig. 11.7 *To avoid the problems shown in Figs. 11.5 and 11.6, large bearings need special design, as shown here, for example, with increased clearance outside the hydrodynamic functionality area.*

11.7 Observation during Operation

During commissioning, the temperature stability should be monitored, and during normal operation, continuous temperature control is highly recommended. Problems with the bearing can change the hydrodynamic lubrication conditions, resulting in a temperature increase. With increasing temperature, the oil viscosity decreases and boundary lubrication is more likely. Boundary lubrication means wear and further increases in temperature. These influences may build up the temperature level very quickly.

As soon as this happens, the machine must be stopped to avoid damage progression. In such a case, if hydrostatic equipment is installed, it should be switched on immediately. The additional oil delays the increase in temperature and reduces the boundary lubrication problem. It may be possible to stabilize the situation by this action but must not lead to the conclusion that the problem has been solved. It is essential to discover the reason for the incident. When the reason is known, then provisions can be made to avoid such an incident in the future.

Temperature control gives early warning of an incident on a bearing. When temperature suddenly increases, any available hydrostatic oil supply should be switched on immediately. This action may delay the damage progress.

Chapter 12

Plain Bearing Damage

Bearing damage was mentioned a number of times in the previous chapter during discussion of assembly and monitoring of plain bearings. Actually, most damages arise from assembly faults or from operational trouble. Damage resulting from design or fabrication faults is rather rare. There is a further category of bearing damages: the failure at the end of the scheduled service life when the fatigue limit is reached. The life expectancy of a plain bearing depends on the application. Automotive bearings experience extremely high loading, and yet the bearings last the life of the engine. At first this seems to be impressive, but with the typical life of a car engine being 3000 working hours, it is a low performance compared with industrial machines. This type of bearing experiences permanent wear from the beginning.

For marine diesel engines, the inspection interval, with possible bearing renewal, is typically after 20,000 working hours. For steam turbines, the intervals are in the region of 50,000 working hours. Hydro power plants may operate for up to 50 years before a bearing change is needed. The wide range of life expectancy reflects the type and the level of loading.

Generally, it can be said that the bearing life improves when the hydrodynamic lubrication conditions are fulfilled and wear is reduced. Design and selection of lining materials are also influential.

Damage at the end of expected bearing life is not spectacular, and usually no extensive damage analysis is called for.

Whether discussing a relatively short-lived car bearing or a long-lasting turbine bearing, unexpected damage always needs a thorough analysis to find the cause. When found, it is usually not difficult to take actions that avoid repeating the same damage situation.

12.1 Preconditions for Long Service

Good basic parameters provide the foundation for long availability of the bearing. They avoid damage or at least delay it for a long time. A wide range of influences must be considered such as calculation, design, selection of material, heat treatment, professional fabrication of the bearings, reasonable quality control, accurate assembly, and finally monitoring of the bearings during operation.

This book has presented a vast accumulation of knowledge on plain bearings, and many hints are given on details and interactions to support a good result. When considering all this, as a start it could be assumed that a plain bearing works totally hydrodynamically and free of any wear. The practice proves that, to the contrary, the life of a bearing always ends with damage.

Then it is necessary to deal with analysis of the damage, and for this it is usual for several groups of people to participate:

- The machine operators who discovered the damage

- The machine supplier

- The bearing producer

- The expert in damage analysis

- The insurer of the machine

Too frequently misunderstandings arise between the parties and disrupt the damage analysis, and sometimes even completely prevent a coherent damage analysis.

The machine operator often does not have detailed knowledge of damage types and therefore asks for an expert. Unfortunately, there are many self-styled experts that do not have the detailed knowledge of plain bearings as presented in this book, and their conclusions are meaningless. A broad expert knowledge is essential for proper damage analysis. Often the terminology causes confusion. If the interactions are not clear and terms become mixed up, the result may be a damage analysis with constant confusion between cause and effect. Such a problem cannot be offset by producing a lengthier report!

If the insurance company is involved in the damage discussion, it can, unfortunately, often be noticed that the amount of valid information from the machine operator is inversely proportional to the extent of the damage. This arises from the fear of losing the insurance cover, but is a misinterpretation of the situation. In fact, all parties have the same priority but for different reasons. The machine operator wants a reliably running machine for good business. The insurer also wants a trouble-free machine with minimum risk of damage claims. Both parties are similarly motivated to find the reason for the damage and to take actions that avoid a repeat of the incident.

Machine operator and insurer both have the same agenda: the machine shall run!

12.2 The Need for Terminology

It is not possible to standardize the subject of bearing damages, but urgently needed is a standardization of clear terminology for bearing damages to allow unambiguous dialogue between the parties.

The DIN 31661, "Plain bearings—Terms, characteristics and causes of damage and changes in appearance," attempts to deal with this topic and states: "The term 'damage to plain bearings' includes all damage and changes in appearance occurring in the bearing surface during operation, whether or not the change or damage adversely affects the performance of the bearing."

This statement is of as little use as the listing of different contact appearances, such as normal wear, edge wear, central wear, etc. The standard contains a large number of photos of damages, some with obscure or incorrect comments. No clear classification by type of damage and damage appearance will be found in this standard, and some important damage types are not mentioned at all. It has often turned out in practice that the application of this standard leads to a misinterpretation of the situation.

The author of this book has been intensively involved with damage analysis for 20 years and very soon learned the essential need to work systematically and in a structured way because of the high complexity of the topic. Excellent basic work has been done by Dr. Kreutzer [16]. As an extension to that, the author describes all the damage interactions clearly arranged in a 15×10 matrix. In the interest of clarity, some simplifications had to be made to allow the final document to stay within manageable proportions. For example, the damage type "contamination" merges everything that does not belong in a plain bearing. This may be all kind of particles, chemicals, or just water in the oil.

Matrix-based damage analysis has been well proven over many years. It cannot replace expert knowledge, but it allows all parties to clearly understand all terms and interactions. That is a basic condition for satisfactory discussion and action about plain bearing damages. The procedure is explained here, beginning with the terminology.

12.2.1 Terminology

For damage analysis, the following terms and definitions should be strictly applied:

12.2.1.1 Damage

Damage is the phenomenon that adversely changes the tribological function of a bearing, usually accompanied by a change in appearance. The damage is initiated by the damage cause and develops to the end of the service life.

The term damage describes the total damage occasion.

12.2.1.2 Damage Cause

The damage cause is the particular event that initiates and leads to a damage.

The damage cause is the initiation of the damage.

12.2.1.3 Damage Appearances

The damage appearance is a defined visible effect on the bearing surface and/ or on the bearing back. A plain bearing failure may show various damage appearances. Usually damage appearances are directly associated with damage characteristics. Typical combinations of damage appearances correspond with a damage characteristic.

A damage appearance is the defined visible effect on the bearing surface.

There are 15 damage appearances defined, clearly different from each other:

- Deposits

- Creep deformation

- Deformation due to temperature cycles

- Thermal cracks

- Fatigue cracks

- Material relief (raised material, lifting of layer)

- Frictional corrosion

- Melting out, seizure

- Polishing, scoring

- Traces of mixed lubrication (intermittent metal to metal contact)

- Blue, black color

- Corrosion, fluid erosion

- Embedded particles, particle migration tracks, formation of wire wool

- Electric arc craters (pitting by sparks)

- Cavitation erosion appearance: material worn out

12.2.1.4 Damage Characteristics

A damage characteristic is the defined description of what has happened in a damage incident based on a detected typical combination of damage appearances. The damage characteristics provide the basis for establishing the cause.

The damage characteristic describes what happened.

There are ten damage characteristics defined, clearly different from each other:

- Static overload

- Dynamic overload

- Wear by friction

- Overheating
- Insufficient lubrication (starvation)
- Contamination
- Cavitation erosion
- Electrical erosion
- Hydrogen diffusion
- Bond failure

12.2.2 The Structure of Damages

The aim of damage analysis is always to find the damage cause. Contrary to popular belief, the damage of a bearing generally gives no direct indication of the damage cause.

To clarify this, here are two examples:

If on a bearing, raised material is detected, then often loss of bond is suggested as the cause. That is wrong! The visible material relief is a typical damage appearance and can be due to any fault during manufacturing, such as wrong preparation or low temperature, and is not a sure sign of loss of bond.

The raised material may also be based on overloading of the bearing. In this case, the cause for overloading has to be found. It could, for instance, be from parts broken off and creating imbalance, leading to the bearing overloading.

These examples show clearly that the cause is not usually visible in the bearing but has to be found by looking beyond the bearing.

The first stage of a qualified damage analysis is to record everything that is visible on the bearing. These are damage appearances. Typical combinations of damage appearances can be tied to a damage characteristic. Having determined the damage characteristic, the damage analysis on the bearing is finished.

The damage characteristic, when established, provides the basis for tracing the damage cause.

The search for the damage cause goes beyond the bearing and needs the support of the operating staff.

Simply to put a damaged bearing on the expert's desk with a request to search for the cause of damage will never be successful. Further detailed information about the damage incident is needed. The story of a damage event is comparable with a pyramid (see Fig. 12.1): at the base there are a multitude of damage appearances. They lead at the next level to a smaller number of damage characteristics. The damage characteristics show the way to the top of the pyramid (i.e., to the damage cause).

Often, it is assumed that a single visible damage appearance provides direct information about the cause of the damage. In practice, that is not so. A failure in progress develops various secondary damage characteristics, each with corresponding damage appearances. As the damage becomes worse over time, the evaluation and search for the cause becomes more difficult. Different causes can lead to identical damage appearances, and one cause can produce different damage appearances.

For accurate damage analysis, all visible damage appearances have to be sorted into primary and secondary categories and related in each case to the associated damage characteristic. Figure 12.2 shows that damage appearances alter with the progress from primary to secondary appearances. If the primary damage characteristic has been found, the search for the damage reason can be started. Because of the obviously complex relationships, there is a need for considerable experience. Equally important is a well-structured procedure based on clear terminology.

Bearing damage has a cause that initiates an effect. The damage gives notice by:

- Increase of service temperature

- Decline of lubrication pressure

- Noise and vibration

- Odor

When these effects appear, the operational reliability of the machine is in doubt and it has to be stopped. A damage analysis then is required to find the cause. As long as none of the mentioned indications are observed, it can be assumed that

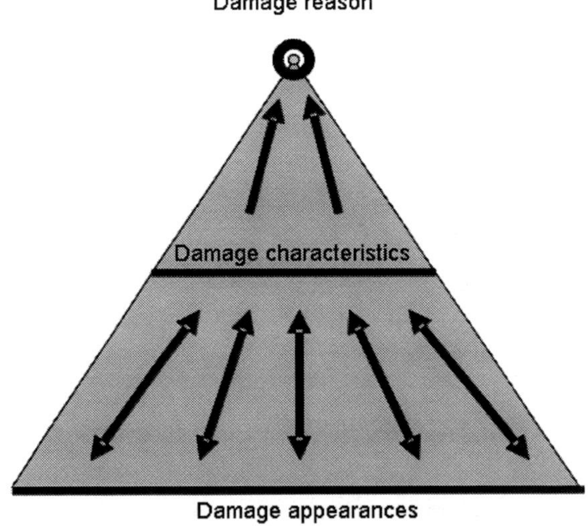

Fig. 12.1 Damage characteristics provide the basis for analyzing the cause.

Damage cause

Primary damage characteristic - damage appearance(s)

Secondary damage characteristic - damage appearance(s)

Secondary damage characteristic - damage appearance(s)

***Fig. 12.2** Damage appearances alter with the progress from primary to secondary appearances.*

the bearings are working correctly. Without clear indications of possible damage in progress, bearings should not be opened and removed just for checking.

Damages can start earlier or later and have the same damage appearances, but for the investigation some clue as to whether the damage happened after short or long service life is very important.

Typical causes of damage after short service life are faults in geometry, assembly, or dirt. In most cases the effects are based on previous damage, and sometimes service conditions modified since last start up are responsible.

Damages after a very long service life usually make a damage analysis unnecessary. Only when there are obviously modified operational conditions, it might be helpful to establish the damage process to avoid similar future incidents.

12.2.3 Damage Appearances

12.2.3.1 Damage Appearance: Deposits

There are various types of deposits. Mineral oil heated above 140°C ages and quickly decomposes into oxides and carbon. These are deposited in the area of highest temperature and show as a brown to black surface discoloration (see Fig. 12.3).

Zinc-based oil additives together with water in the oil in contact with a copper oil cooler result in decomposition. Oil aging rapidly follows, and a red-brown deposition in the area of maximum temperature will be seen. This deposit can reach a significant thickness and may finally block the oil filter.

A deposit known as "wire wool" can also affect the functionality of a bearing. "Wire wool" develops when hard particles embedded in the bearing cut the shaft. This risk is noticeable when using shaft material with chromium content.

12.2.3.2 Damage Appearance: Creep Deformation

Creep is the crack-free plastic deformation of the bearing lining in the direction of rotation. Creep deformation is triggered by high pressure and aggravated by high temperature. White metals are liable to this effect unless they are hardened by special elements.

The tongue-shaped surface deformation produced by creep is seldom seen in original condition because the bulge reduces clearance, inhibits lubrication, and wear immediately results. Bearings with intensive creep deformation show material squeezed out at the edges of the lining thickness, as shown in Fig. 12.4.

12.2.3.3 Damage Appearance: Deformation Due to Temperature Cycles

On rapid machine start-up, where very high temperature is reached within a short time, tin-based bearing linings may develop increased surface roughness. At first the surface structure has an orange peel appearance. Following extreme temperature cycles, with many machine starts, a columnar crystal structure is formed; see Fig. 12.5.

12.2.3.4 Damage Appearance: Thermal Cracks

Thermal cycling at high temperatures may influence the tin-based layer adversely when the heterogeneous material consisting of a relatively soft matrix and embedded hard crystals cannot withstand the cycles. Thermal cracks form along grain boundaries. The effect is more intense when the tin-based layer is contaminated by lead. Unlike cracks caused by material fatigue, the thermal

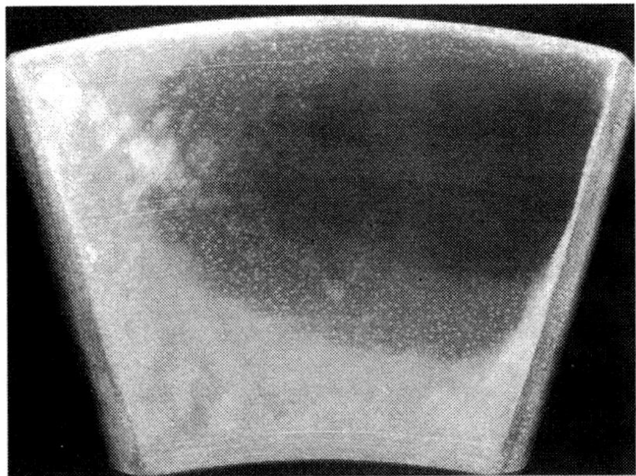

Fig. 12.3 Deposit of oil carbon on the pad of a thrust bearing.

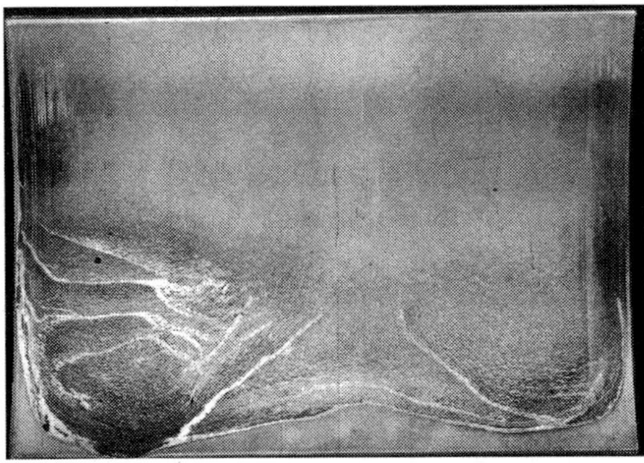

Fig. 12.4 Creep deformation of the lining. At bottom left, the distortion of the lining border is clearly visible protruding beyond the bearing.

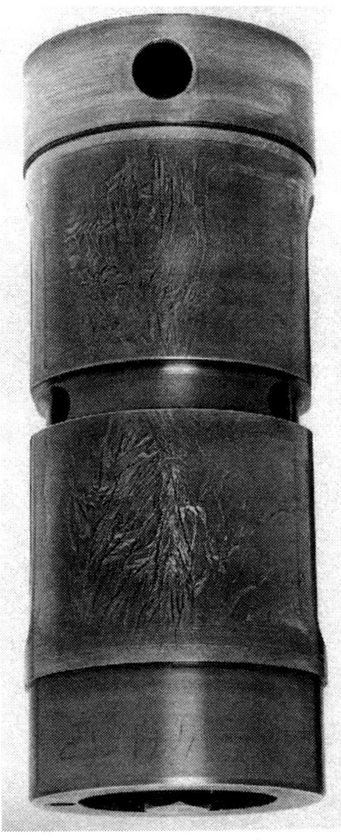

Fig. 12.5 Columnar crystal surface structure following extreme and frequent temperature cycles.

cracks follow a shaky, disorientated path with pit mark outbreaks. Figure 12.6 illustrates an early stage of thermal cracking. In advanced stages, the damage is very similar in appearance to cavitation erosion. The difference is that thermal cracks are always in combination with other damage appearances.

Fig. 12.6 *Thermal cracks at an early stage.*

12.2.3.5 Damage Appearance: Fatigue Cracks

Cracks occur as a result of exceeding the material fatigue strength. They start on the lining surface, continue vertically through the lining to the bond interface, and follow the bond line. Finally, they meet adjacent cracks, and the result is delamination of lining areas, known as crazy paving (see Figs. 12.7 and 12.8). The loose parts remain in their positions until the bearing is opened. After opening of the bearing, the lining pieces usually adhere to the oil on the shaft. The appearance of material fatigue in the first instance seems to be similar to that of thermal cracks or cavitation erosion. Precise inspection always clearly shows the border of the fatigue area forming a radius from the bond interface along the lining thickness. It has a ground appearance.

12.2.3.6 Damage Appearance: Material Relief (Raised Material, Lifting of Layer)

There are three possibilities for raised material. The previously mentioned fatigue ends with raised lining material. Raised material can also be a result of blistering. When it happens with new backings above 60 mm thick, then it is based on problems with hydrogen diffusion (shown in Fig. 12.9).

Blistering also occurs on thermal sprayed layers; see Fig. 12.10. This has to be metallographically investigated. In the broader sense, it is also a loss of bond, a bond defect.

Low bond strength and bond defects also result in raised material, either in one large area or many small areas along the bond surface, as shown in Fig. 12.11. In the defect areas, surface corrosion is always visible on the backing when the backing preparation has been done incorrectly. There may also be dark lines visible from hot spots generated during the casting process.

Fig. 12.7 *Principle of fatigue of the lining in progress from the left to the right.*

Fig. 12.8 *Fatigue with typical "crazy paving" delamination.*

12.2.3.7 Damage Appearance: Frictional Corrosion

Frictional corrosion can be seen on the contact surfaces of backing and housing when they do not fit well. Very small relative movements occur and result in local seizure effects that show as dark colors on the surfaces. Later, as damage progresses, small weld puddles also become visible. They look similar to, and will be often be confused with, those from current flow. Usually frictional corrosion on the bearing outside (Fig. 12.12) will occur, together with raised lining material due to fatigue of the layer.

12.2.3.8 Damage Appearance: Melting Out

When melting out occurs, it becomes clearly visible on the running surface; see Fig. 12.13. On journal bearings, the molten material solidifies and is wiped onto the lining surface in the direction of rotation. If there is any doubt, metallographic investigation will show that the original lining and the wiped layer are clearly different in crystallization structure (see Fig. 12.14).

12.2.3.9 Damage Appearance: Polishing, Scoring

After only a few machine starts, the effects become visible in the area of maximum load on the white metal lining. This results from short periods of mixed lubrication conditions during each start and stop. In principle, this is wear, but in practice it is no more than a surface smoothing by burnishing. The affected surface area is polished compared with the other areas, but there is no adverse effect for the bearing's operation.

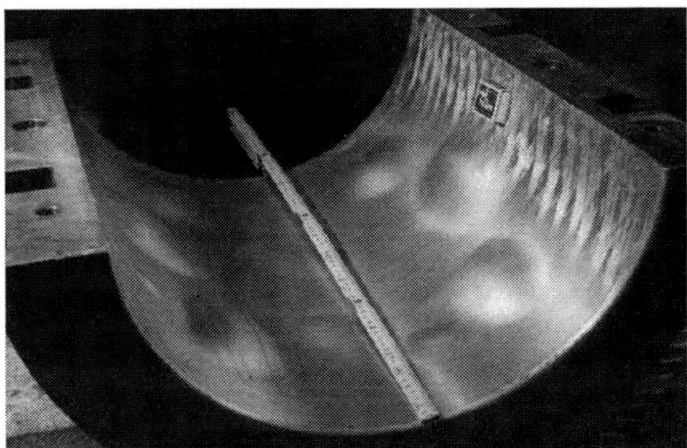

Fig. 12.9 *Hydrogen diffusion blisters forming raised material.*

Fig. 12.10 *Blister forming raised material in the oil pocket. The lining has been made by thermal spraying.*

Fig. 12.11 *Material raised by loss of bond. Corrosion is visible on the dark areas of bond surface with frictional corrosion to the backing.*

If a journal bearing is assembled out of alignment, then the polished area is asymmetric. The degree of asymmetry and eccentricity supplies clues about the severity of misalignment.

During the polishing contact, small particles may produce scoring scratches in the direction of rotation, as shown in Fig. 12.15. Single scratches with little depth are not harmful to the bearing operation.

12.2.3.10 Damage Appearance: Signs of Mixed Lubrication, Surface Wear

Wear is documented by traces of mixed or partial lubrication, whereby the location gives further hints:

When a journal bearing shows clear signs of mixed lubrication at one end of the running surface, then the bearing is misaligned to the shaft.

Intensive traces uniformly distributed in the maximum load zone are the result of prolonged running with a reduced lubrication gap. The cause may be high load, high operation temperature, reduced oil supply, inadequate oil viscosity, or even contamination with particles.

Traces of mixed lubrication away from the loaded area give a clear indication of deviation of geometry; see Fig. 12.16.

Fig. 12.12 Frictional corrosion on the outside of the bearing.

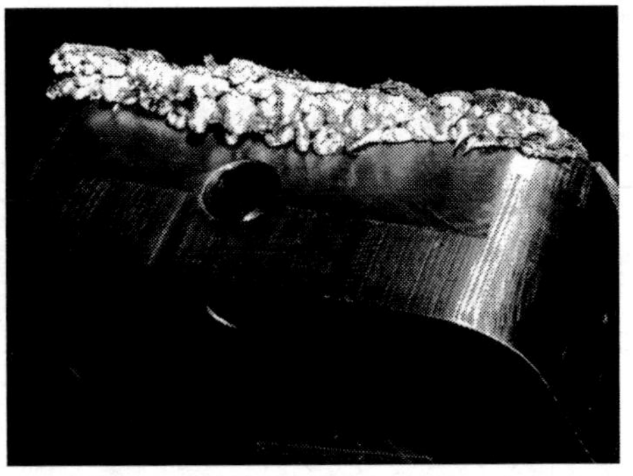

Fig. 12.13 White metal lining clearly melted away from the pad.

12.2.3.11 Damage Appearance: Blue, Black Color

Subject a bearing to temperatures above 300°C, and the steel backing takes on a blue annealing colour and the lubricant dries to a black film. This cannot be clearly shown with black-white photographs; therefore, none is included here.

12.2.3.12 Damage Appearance: Corrosion, Fluid Erosion

Chemicals circulated through the bearing with the lubricant can activate corrosion on the bearing surface layer. The fluid flow supports the damage progress, and in an advanced stage more erosion cavities become visible. The appearance of this roughened surface can easily be confused with electric arc pitting. Metallographic investigation can clarify. See Figures 12.17 and 12.18.

12.2.3.13 Damage Appearance: Embedded Particles, Particle Migration Tracks

Particles embedded in the lining surface in most cases are the cause of score marks in the direction of movement. Metallographic investigation provides information about the type of particles. Figures 12.19 and 12.20 illustrate embedded non-metallic and metallic particles.

12.2.3.14 Damage Appearance: Electric Arc Craters (Spark Pitting)

On electrical machines, magnetic fields can lead to voltage differences between journal and bearing. The electric current may bridge the lubrication gap; arcing occurs, and arc craters are melted into the lining and journal surfaces, as shown in Fig. 12.21. The affected areas are aligned in the direction of running and are rough. Micrographic investigation clearly shows the arc craters; see Fig. 12.22.

12.2.3.15 Damage Appearance: Cavitation Erosion

Cavitation occurring in lubricants leads to implosions. The developing low pressure results in a local breakout of very small particles from the layer surface; see Fig. 12.23. More and more layer material becomes removed as the damage progresses. The appearance seems to be similar to that of thermal cracks but with the difference that cavitation erosion shows no other damage signs.

Fig. 12.14 *The surface of the lining has been molten. Clearly visible is the changed crystallization structure.*

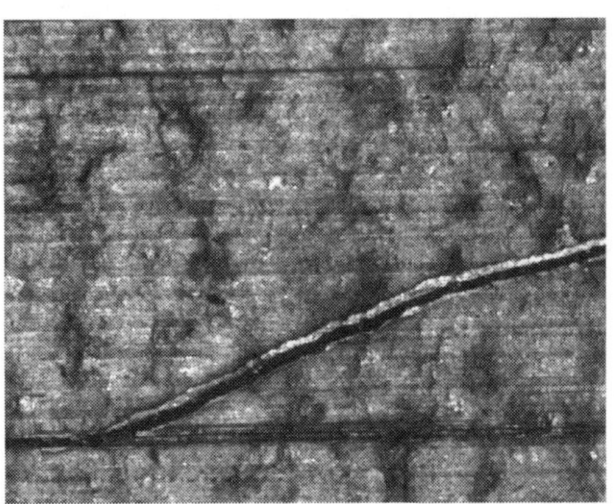

Fig. 12.15 *Scoring in direction of bearing motion.*

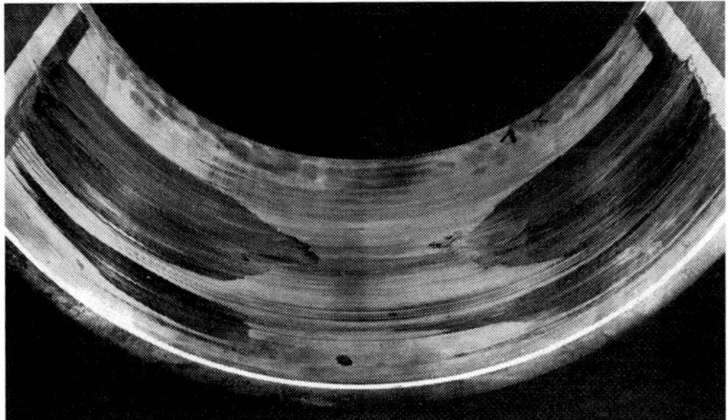

Fig. 12.16 *Traces of mixed lubrication outside the loaded area giving a clear hint to geometry deviations.*

12.2.4 Damage Characteristics

12.2.4.1 Damage Characteristic: Static Overload

The corresponding damage appearances are creep deformation, evidence of mixed lubrication with some wear, deposits, and thermal cracks.

The bearing is loaded beyond its capacity, especially when starts are frequent or under preloading. Unexpectedly high temperatures will further reduce the loading capacity. With static overload, the lubrication clearance is reduced, which causes a temperature rise. Creep deformation and mixed lubrication may occur. The worn material is wiped back onto the lining surface further along. Deposits of oil carbon and thermal cracks may follow; see Fig. 12.24.

12.2.4.2 Damage Characteristic: Dynamic Overload

The corresponding damage appearances are fatigue cracks, raised material, and frictional corrosion between the bearing outside diameter and the housing. See Fig. 12.25.

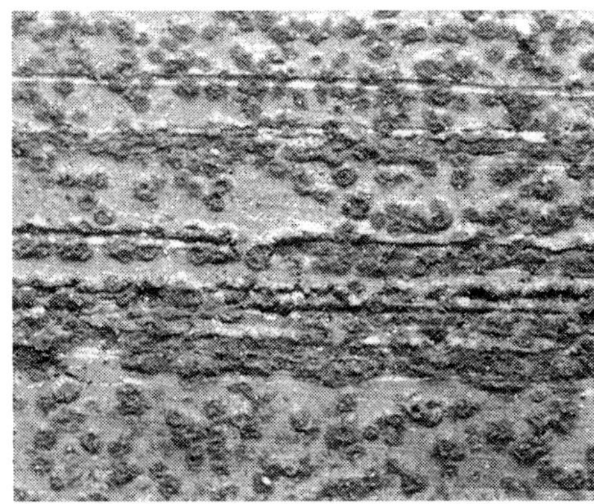

Fig. 12.17 Corrosion on a plain bearing.

Fig. 12.18 Metallographic investigation clearly shows the corrosion on the surface.

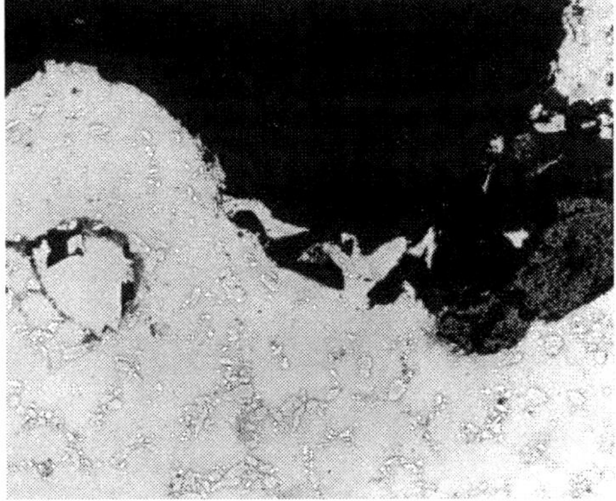

Fig. 12.19 Embedded non-metallic particles.

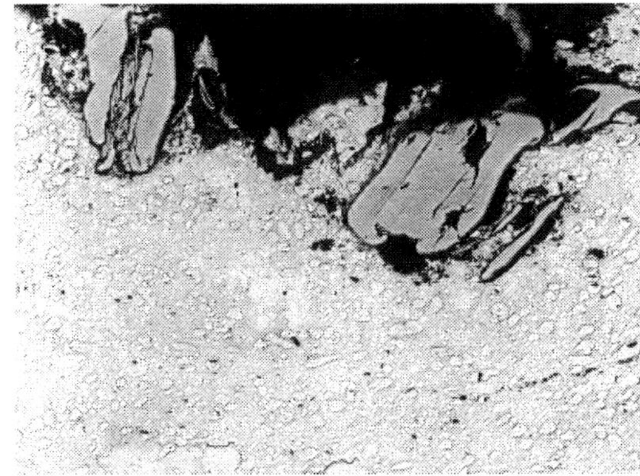

Fig. 12.20 Embedded metallic particles.

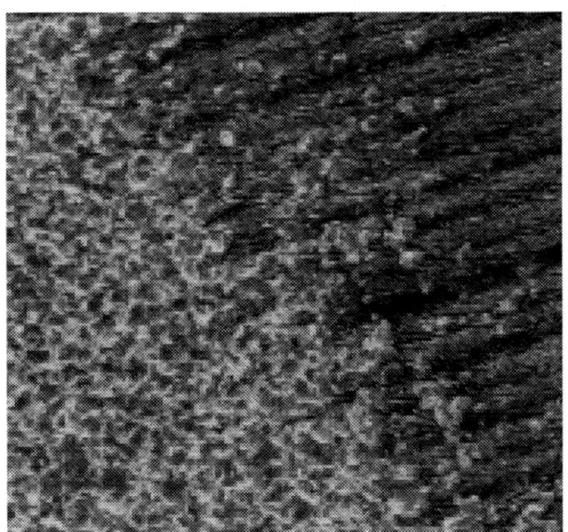

Fig. 12.21 Electrical pitting. Transition to undamaged surface. On the undamaged area, the machining scores are clearly visible.

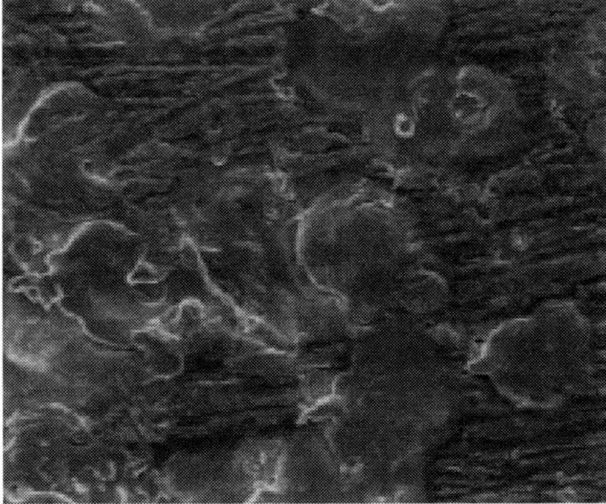

Fig. 12.22 The micrographic investigation clearly shows the craters.

The bearing has been run above the limit of its dynamic load capacity. Abnormally high temperature levels further reduce the loading capacity. Flimsy design of bearing shell and housing leads to increased local bending deformation of the lining and shell, which reduces the fatigue strength. If the bearing is a loose fit in its housing, then the same symptoms can occur. In such a case, the frictional corrosion is more pronounced and clearly visible. Dynamic overload can be precisely identified by the rounded borders of the area of the raised lining layer.

Often it is assumed that the raised lining material is a result of loss of bond, because beneath it there is no tinning visible on the steel surface. This assumption is wrong. The "crazy paving" pieces of lining, detached by fatigue cracks, are agitated by the lubrication stream and polish the bond surface down to clean steel.

12.2.4.3 Damage Characteristic: Wear by Friction

The corresponding damage appearances are polishing, scoring, and indications of mixed lubrication (some metal-to-metal contact); see Fig. 12.26.

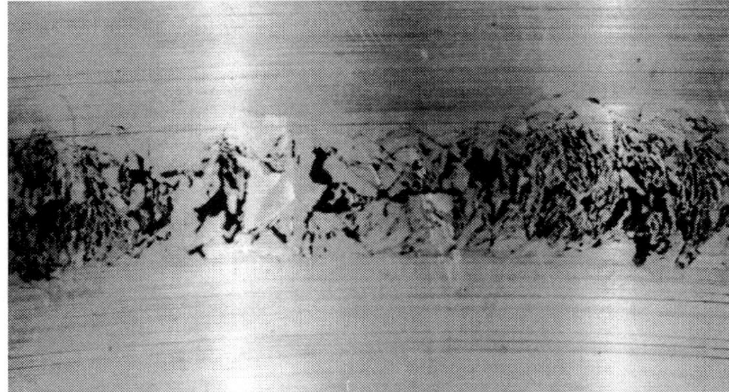

Fig. 12.23 *At the start of cavitation erosion, tiny particles are broken out of the layer surface.*

Fig. 12.24 *Static overload due to tight assembly.*

Frequent starting and stopping means frequent mixed lubrication conditions, and low rotation speed combined with high loading may extend the time these conditions apply. Further causes of wear by friction are deviations of geometry due to machining tolerances, shaft deformation, housing deformation, and assembly faults.

12.2.4.4 Damage Characteristic: Overheating

The corresponding damage appearances are deposits, creep deformation, deformation due to temperature cycles, thermal cracks, and mixed lubrication; see Fig. 12.27.

Most of the heat generated in the bearing is carried away by the lubricant. The remainder goes by thermal conduction and to a lesser extent thermal radiation. If the heat is not removed effectively, overheating will occur.

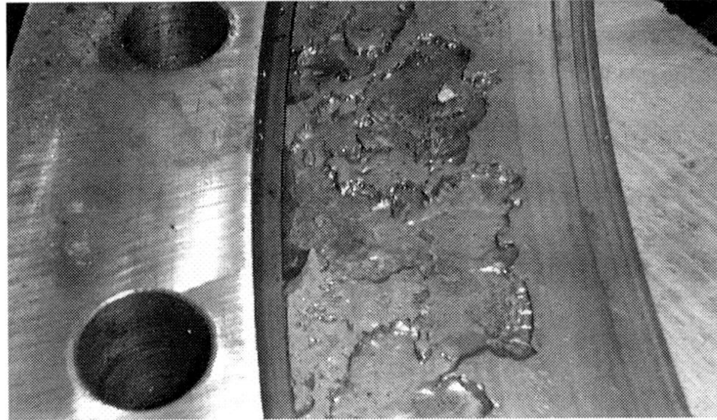

Fig. 12.25 Dynamic overload: after removing the lifted lining material along the border of the damaged area, the typical transition radius in the layer becomes visible. It is a sure sign of a dynamically overloaded bearing.

Fig. 12.26 Frictional wear on a thrust bearing. The worn material from the adjacent pad has been deposited along the inlet radius on the left side. The lubrication feed to the pad was hindered and the wear increased.

12.2.4.5 Damage Characteristic: Insufficient Lubrication (Oil Starvation)

The corresponding damage appearances are traces of mixed lubrication, melting of the lining, and blue and black discoloration, as shown in Fig. 12.28.

The available supply of lubricant is not sufficient to allow hydrodynamic running. The temperature steadily increases, the lubrication deteriorates, and finally the bearing runs nearly dry.

12.2.4.6 Damage Characteristic: Contamination

The corresponding damage appearances are deposits, scoring, signs of mixed lubrication, corrosion, erosion, and embedded particles. See Fig. 12.29.

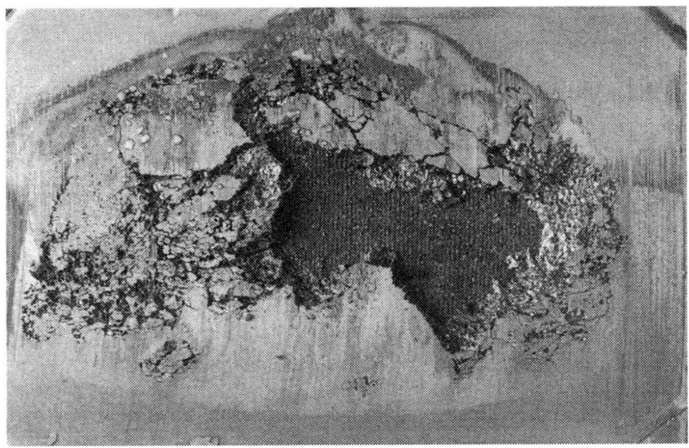

Fig. 12.27 Overheating: on the outer area creep deformation is clearly visible, and in the mid range there are thermal cracks.

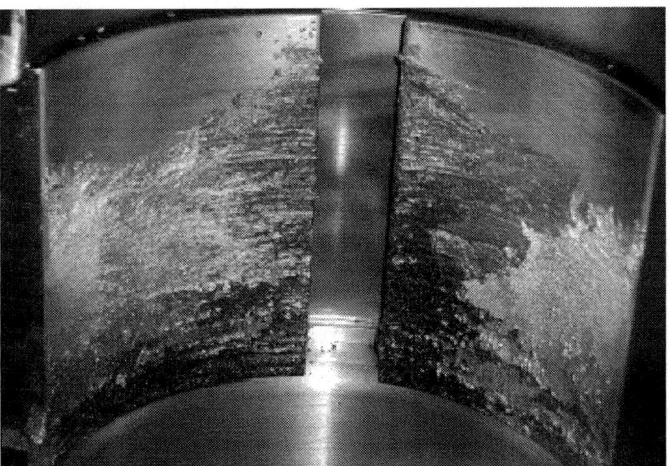

Fig. 12.28 Insufficient lubrication led to wear, followed by intensive melting and wiping. Due to high temperature, the surface is blackened by oil carbon.

The term contamination covers all substances in the lubricant that are not intended to be present in the system and will be divided into two groups: foreign particles and chemicals. They may enter the machine before commissioning, in which case the cause is carelessness. Typical are drilling swarf, welding debris, grinding residues, cleaning rags, foundry sand, blasting grit, and oxide scale.

Fig. 12.29 Cement powder entered the lubrication circuit and damaged the bearing of a cement mill.

Particles may also be a result of machine operation. Typical are wear particles from bearings, shafts, gear teeth, piston rings, seals, and flaking paint.

When seals are defective, contamination particles can originate from outside the machine, especially when processing granular materials such as sand or cement.

Chemical contaminations may, for example, be acid deposits from engine combustion or ammonia from ammonia gas compressors.

Wire-wool contaminant develops when using shafts of austenitic steel containing chromium. Hard oxide particles become embedded into the white metal bearing surface, and they cut steel threads from the shaft. High sliding speeds aggravate the effect. The wire wool is circulated by the lubricant and will block the filters.

12.2.4.7 Damage Characteristic: Cavitation Erosion

The corresponding damage appearance is erosion and is unique; see Fig. 12.30.

In contrast, the similar looking thermal cracks are always attended by other damage appearances, and the damage appearance of fatigue cracks shows the typical transition radius along the layer border.

If well-proven machines suddenly develop cavitation problems, then invariably a water content of more than 0.2% in the oil is responsible. Damage is typically found in the unloaded bearing area.

12.2.4.8 Damage Characteristic: Electrical Erosion

The corresponding damage appearances are electric arc craters, shown in Fig. 12.31.

The electric arc craters are the only visual sign, but they characteristically are found both on the bearing surface and the shaft.

On electric machines, magnetic fields can stray and induce an electric current flow to ground via shaft and bearing in the absence of a proper grounding connection.

12.2.4.9 Damage Characteristic: Hydrogen Diffusion

The corresponding damage appearance is lifting of the bearing material layer, shown in Fig. 12.32.

Fig. 12.30 *Material lifting due to cavitation erosion.*

Fig. 12.31 *The left half of the pad experienced intensive electric current flow, which caused pitting by electric arc craters.*

When the bearing material lifts into blisters on new bearings that have not been lined by metal spraying, then it is always based on hydrogen diffusion (see Section 3.4.3).

12.2.4.10 Damage Characteristic: Bond Failure

The corresponding damage appearance is lifting of the bearing material layer.

Along borders and edges of joint faces, there is the risk of falling below the recommended bearing alloy casting temperature during manufacture. Bond failure follows, but in most cases it is very local and usually can be restored by qualified treatment with a soldering procedure.

The situation is different when ultrasonic testing shows a multitude of small bond defects distributed over the entire bearing. This is a clear indication of weak bond strength in general. Often there is the risk of these defects growing in size during handling, transportation, or in service. Large unbonded areas are formed.

Low bond strength can be detected quite simply with a chisel test. If the lining can be removed easily without much force, then clearly the bond strength is very low.

12.3 Analysis Using the Damage Matrix

Although strictly speaking there is a difference between damage appearance and damage characteristic, for most people the subject remains confusing.

Fig. 12.32 Hydrogen diffusion on a thick-walled bearing.

The more so as some damage appearances can correspond with different damage characteristics. Furthermore, there are primary and secondary damage appearances with corresponding damage characteristics. For clarification, the author has developed a damage matrix, shown in Table 12.1. Here the 15 damage appearances and the 10 damage characteristics are listed, and the interaction is clearly shown.

An example: the damage appearance mixed (boundary) lubrication corresponds with five damage characteristics. The damage appearances of deformation due to temperature cycles, fatigue cracks, frictional corrosion, melting out, corrosion, embedding of particles, cavitation erosion, and electric arc craters (pitting) each correspond with only one damage characteristic. A specific feature is the damage appearance of material lifting. It corresponds to static overload, hydrogen diffusion, and bond failure, but in each of the three cases the visible appearance is clearly different: the dynamic overload leads to typical crazy paving delamination; the hydrogen diffusion leads to blister delamination; and finally the bond failure delamination can easily be confirmed by chisel test.

When all visible appearances of a damage incident are listed in the matrix, then matching with the corresponding damage characteristics becomes very easy. If

Table 12.1 The Damage Matrix

Deposits	Creep deformation	Deformation due to temperature cycles	Thermal cracks	Fatigue cracks	Material relief	Friction corrosion	Melting out, seizure	Polishing, scoring	Traces of mixed lubrication	Blue, back color	Corrosion, fluid erosion	Embedded particles	Electric arc craters	Cavitation erosion appearance: material worn out	Damage characteristics
•	•		•						•						Static overload
				•	•	•									Dynamic overload
								•	•						Wear by friction
•	•	•	•						•						Overheating
							•		•	•					Insufficient lubrication
•								•	•		•	•			Contaminations
														•	Cavitation erosion
													•		Electro erosion
					•										Hydrogen diffusion
					•										Bond failure

the appearances lead to several damage characteristics, then both primary and secondary characteristics have been detected.

12.3.1 Examples Using the Damage Matrix

Figure 12.33 shows a damaged bearing with the appearance of fatigue cracks (typical crazy paving cracks). Lifting of the lining is also visible, but not in blisters, and a chisel test confirms good bond in the surrounding area. The lifting material therefore clearly corresponds with the damage characteristic of dynamic overload. Very often (but not invariably) with dynamic overload, frictional corrosion will be found on the outside of the bearing. The three appearances found—thermal cracks, lifting material, and possibly frictional corrosion—lead clearly to the characteristic of dynamic overload. With this result, the damage analysis ends. The next step is to involve the machine operator in the investigation of why there was a dynamic overload: if frictional corrosion was visible, the first task is to check the fit between bearing and housing for tolerances and roundness. Is the rigidity of the bearing and housing assembly sufficient? Or perhaps was there a problem with unbalanced masses?

Figure 12.34 shows a bearing damage that leads to four characteristics from the four appearances of creep deformation, thermal cracks, mixed lubrication, and discoloring.

Creep, thermal cracks, and mixed lubrication correspond to the characteristic of static overload.

Mixed lubrication corresponds also to the characteristic of wear.

Creep, thermal cracks and mixed lubrication lead to overheating.

Mixed lubrication and discolored surface point to the characteristic insufficient lubricant.

Mixed lubrication corresponds also to the characteristic of contamination, but this can be excluded because it fulfills only one of five possible hits.

The damage characteristics match static overload, wear by friction, overheating, and insufficient lubrication. In most cases, the insufficient lubrication is secondary and consequential to the damage characteristic of overheating. Wear by friction follows as secondary to the static overload. Therefore, the search for the damage reason should be concentrated on the damage characteristics of static overload and overheating: for example, was there an unusually high radial load? Was the bearing clearance too little? Was the heat flow hindered? Was start up too quick? Was the rotation speed much higher than usual?

This example shows that a damage analysis can show multiple damage characteristics. The more damage appearances that relate to one damage characteristic, the more likely is that characteristic to be the cause of the damage.

Finally, it must be established which of the damage characteristics is primary and which is secondary. If there is any doubt about this, all the damage characteristics found have to be considered as primary.

We do not expect that everybody can produce an exact damage analysis by following these instructions. The damage matrix simply supplies transparency of definitions and shows the interactions so that everybody involved in a damage investigation can follow the discussion without ambiguity.

The author developed the terminology and the damage matrix presented here as a project contributing to international standard ISO 7146. The damage

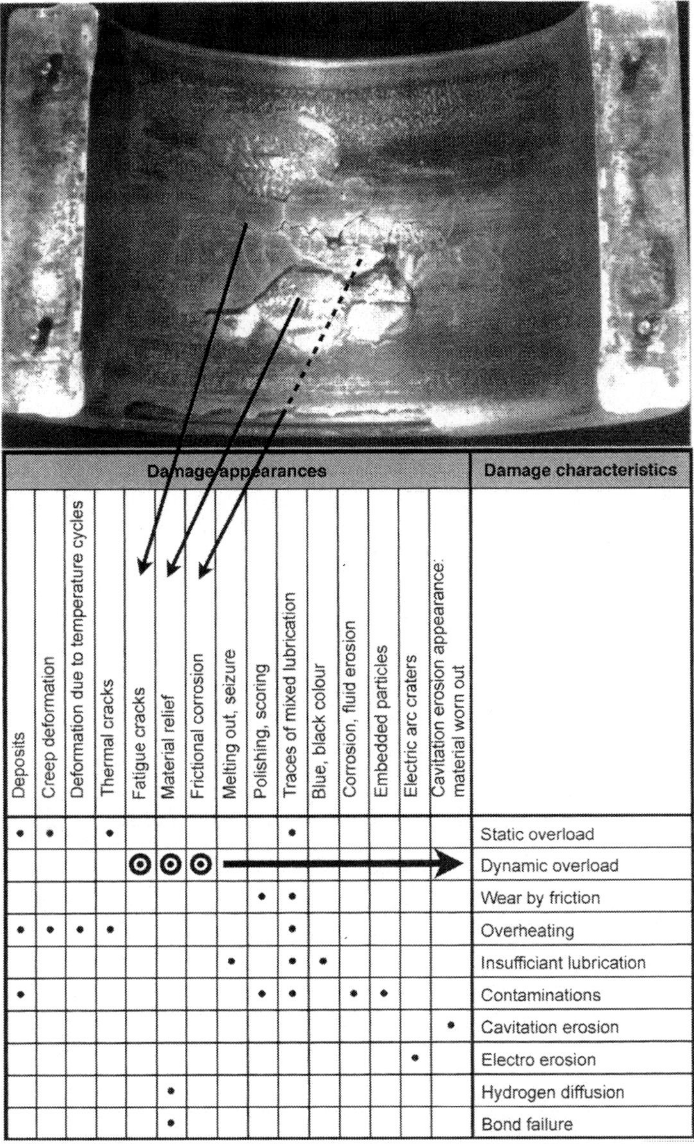

Deposits	Creep deformation	Deformation due to temperature cycles	Thermal cracks	Fatigue cracks	Material relief	Frictional corrosion	Melting out, seizure	Polishing, scoring	Traces of mixed lubrication	Blue, black colour	Corrosion, fluid erosion	Embedded particles	Electric arc craters	Cavitation erosion appearance: material worn out	Damage characteristics
•	•		•						•						Static overload
				⊙	⊙	⊙									Dynamic overload
								•	•						Wear by friction
•	•	•	•						•						Overheating
							•		•	•					Insufficiant lubrication
•								•	•		•	•			Contaminations
														•	Cavitation erosion
													•		Electro erosion
					•										Hydrogen diffusion
					•										Bond failure

Fig. 12.33 *Example of a damage analysis leading to one damage characteristic.*

characteristic of cavitation erosion is, in this book as well as in ISO 7146 part 1, limited to problems caused by water in the oil. That is the only possible reason for sudden cavitation damage on well-proven machines. However, part 2 of ISO 7146 deals comprehensively with all variations of cavitation damage and will be found helpful for designing new machines.

ISO 7146 part 1 has over fifty pages in which all damage characteristics are discussed in detail. There are also photos of damaged bronze and galvanic bearings. Here in this book the focus is more on the damage appearances, and so the two documents complement one another perfectly.

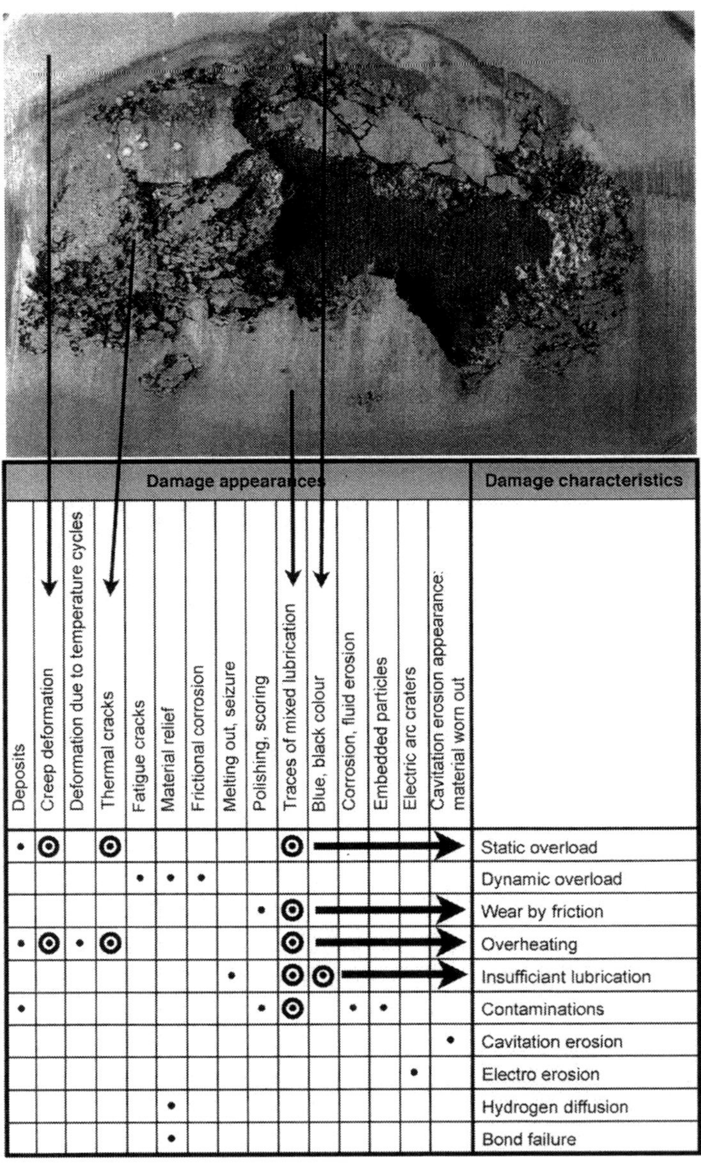

Deposits	Creep deformation	Deformation due to temperature cycles	Thermal cracks	Fatigue cracks	Material relief	Frictional corrosion	Melting out, seizure	Polishing, scoring	Traces of mixed lubrication	Blue, black colour	Corrosion, fluid erosion	Embedded particles	Electric arc craters	Cavitation erosion appearance: material worn out	Damage characteristics
•	◉		◉						◉					→	Static overload
				•	•	•									Dynamic overload
								•	◉					→	Wear by friction
•	◉	•	◉						◉					→	Overheating
							•		◉	◉				→	Insufficiant lubrication
•								•	◉			•	•		Contaminations
														•	Cavitation erosion
													•		Electro erosion
					•										Hydrogen diffusion
					•										Bond failure

Fig. 12.34 Example of a damage analysis leading to multiple damage characteristics.

12.4 Damage Report

In practice, there can be found many damage reports that are more of a disaster than the investigated damage itself. Such reports lead to wrong interpretation, and consequently action meetings follow the wrong direction. The total damage increases.

Adopting the following suggestions will put everybody in the position to check damage analysis reports for their quality. First, the terminology of ISO 7146, and the methods explained there, should be applied. Creators of new and modified terms will not be clearly understood. Typically in such reports, reasons and actions will be continuously exchanged.

Damage reports must use only the terminology of ISO 7146.

Detached lining material is not always automatically a bond defect. A bond defect is not the reason for damage, as often claimed, but a damage characteristic. There are various possible causes of detached lining material, as detailed in this book. To argue that detached material is automatically a bond defect due to faulty lining is in almost all cases wrong. The conclusion is not improved with the detection of Fe, Sn, and machining grooves on the bonding surface. This is pure actionism.

Insufficient lubrication also describes a situation, and thus it is classified a damage characteristic. Every bearing damage instance finally ends in insufficient lubrication when the progress is long enough and, in this case, the insufficient lubrication is secondary.

"Oil Starvation" is a popular term, although it allows no inference of the damage reason.

When loss of bond, insufficient lubrication, or oil starvation is stated as the reason for damage, then scepticism is highly recommended.

Anyone who declares the visible damage appearance or the corresponding damage characteristic as a damage reason is not familiar with the basic rules of a damage analysis.

Experts with practical experience are clearly the best qualified for a damage analysis, and comprehensive knowledge of material science, fabrication of bearings, and operation of bearings is desirable. This approach is the key to successful bearing damage analysis but has the disadvantage that the number of acceptably qualified experts is extremely limited.

To find the reason for a damage incident is in the interest of all concerned if it enables successful provisions to be made to avoid repetition of the same damage. For accurate damage analyses, the scheduled rules and terms have to be strictly followed, otherwise misunderstandings will arise. The damage matrix is a well-proven tool for clearly structuring the damage analysis and minimizing wrong interpretations.

It is highly recommended to avoid making a damage analysis just by comparison of photographs. This will often lead to wrong interpretation, because exposure to the light may change the appearance of a photograph and dramatically affect the evidence it seems to convey.

Chapter 13

The Future of Plain Bearings

This book focuses on problems, inadequacies, and the demands of improvement and research. The reader, especially if unfamiliar with plain bearing technology, may infer that these bearings are an unreliable machinery component. The reality is quite the opposite. Plain bearings are now very safe and reliable, and they should continue to be so in the future.

The author comments on the changes in the plain bearing industry and provides some new information and ideas. The demands on plain bearings are increasing in all areas, from material and manufacture to installation and operation, so in the future, parameters will change and interactions have new priorities. Inadequacies that in the past gave no ill effects, thanks to the good-natured behavior of the plain bearing, may in the future become problems, due to changed and more demanding conditions.

To avoid being caught out by such circumstances, it is never too soon to start appraisal of all influences and to give them serious consideration, even if they seem, initially, to be trivial. Only by such action can we be in position to notice unexpected effects and arising problems and produce early solutions.

The fundamental challenge is to correctly identify all influences.

It is human nature to assume that well-known conditions and rules are eternally valid and use them as the basis for all actions without question. In this book, it has been shown that developments based on such blind faith may lead to a dead end.

We do not suggest that everything has to be taken into question, but a little more scepticism is not a bad thing. There is no sure method to detect all unknown influences systematically. The experts will detect them over time, in many cases even by chance. It is important to collect such experiences and to document them. The summary of this information is an ideal basis for future R&D.

With the new laser lining technique, all restrictions on the material composition are removed, and the way is open for further optimization. At the same time, the new technique gives an unprecedented stability to the process at the highest level. These discoveries have to be optimized in the future for practical use. At the same time, the industry should put aside designs with dovetails and thick lining layers. The material data must provide more useful information, and the nondestructive test procedures for quantitative evaluation of bond strength must be improved. Plain bearing technology is not antiquated, nor are the R&D themes exhausted. The future has already begun, and there is much to do.

In this book, the author has summarized many of his observations. The collection is by no means complete, but it is a beginning and possibly will motivate other experts to complete or add to the themes, to discuss them, and to utilize the results in future actions. If this is achieved, then it was worth the effort of writing it.

References

[1] G. Vogelpohl: Betriebssichere Gleitlager, Erster Band, Springer Verlag, 1967.

[2] O. R. Lang, W. Steinhilper: Gleitlager, Springer Verlag, 1978.

[3] W. Hailers: Gleitlagerwerkstoffe, Gleitlagertechnik, Th. Goldschmidt AG, 1992.

[4] K. Spiegel: Zur Staggering der Belastbarkeit von Gleitlager-Verbundwerkstof-fen, Tribologie + Schmierungstechnik, 52. Jahrgang, Expert Verlag, 2005.

[5] W. Hailers: Gleitlagerermüdung – Bestimmung der dynamischen Festigkeit von Gleitlagerwerkstoffen, Tribologie Band 6, Springer Verlag, 1983.

[6] R. Löhr, D. Eifler, E. Macherauch: Ermüdungsmerkmale bei der Schwingbean-spruchung des Gleitlagerwerkstoffes SnSb8Cu4Cd im Temperaturbereich, 20 °C - 100 °C, Tribologie Band 3, Springer Verlag,1982.

[7] R. Löhr, E. Macherauch, D. Eifler: Grundlagenorientierte Untersuchungen zum Dauerschwingverhalten von Gleitlagerwerkstoffen, Tribologie + Schmierungs teaching, 32. Jahrgang , Expert Verlag, 1985.

[8] R. Löhr, P. Mayr, E. Macherauch: Untrusting zum Ermüdungsverhalten eines hochzinnhaltigen Gleitlagerwerkstoffs, Goldschmidt informiert 3/80.

[9] H. Gröber: Die Grundgesetze der Wärmeleitung und des Wärmeübergangs, Springer Verlag, 1921.

[10] E. Schmidt, R. Weber: Gleitlager, Springer Verlag, 1953.

[11] M. Khonsari, E. Booser: Applied Tribology, John Wiley & Sons, 2001.

[12] G. Knoll: Tragfähigkeit zylindrischer Gleitlager unter elastohydrodynamischen Bedingungen, Diss. TH Aachen, 1974.

[13] H. Peeken, G. Knoll, J. Widyanata: Lagerverformung, Tribologie Band 3, Springer Verlag, 1982.

[14] G. Knoll: Dynamisch belastete Gleitlager im Kurbeltrieb, VDI Wissensforum, 2010.

[15] R. Datta, R. Haller: Gleitlager-Ermüdung, Tribologie 3, Springer Verlag, 1982.

[16] R. Kreutzer: Schäden an Gleitlagern, Allianz Zentrum für Technik GmbH, 1990.

Credits

1   Fig. 2.1, Ennskraftwerke AG, Steyr.

2   Figs. 2.5, 2.7, 2.8, and Cover Photo, John Crane Bearing Technology GmbH, Göttingen.

3   Fig. 2.9, Renk AG, Hannover.

4   Fig. 2.10, Main-Metall Tribologie GmbH, Altenglan.

5   The brands ECKA, HOYT and LACOMET are registered trademarks or trademarks of ECKA Granules Germany GmbH or its affiliated companies in Germany and/or other countries. TEGO, TEGOTENAX, TEGOSTAR and ROPTIN are registered trademarks or trademarks of Evonik-Goldschmidt GmbH in Germany and/or other countries, which are used by ECKA Granules Germany GmbH or its affiliated companies under license. Without prior written consent of the ECKA Granules Germany GmbH you are not allowed to use, copy, reproduce, republish, transfer, distribute or change the above mentioned trademarks in whatever way. ECKA® TEGOSTAR™ patented US 5 705 126, EP 0717121 and worldwide. ECKA® LACOMET® patent pending PCT/EP2006/063159.

6   A part of the work reported in this chapter belongs to the Hercules B research program of the European Commission Framework Program FP7. Partners involved in the work were MAN Diesel A/S, Miba Gleitlager GmbH, and ECKA Granulate GmbH & Co. KG.

7   THRUST computer program from Rotating Machinery and Controls Industrial Research Program (ROMAC), University of Virginia.

All remaining figures: ECKA Granules Germany GmbH.

Index

About the Author
Rolf Koring

Rolf Koring studied mechanical engineering sciences at the University Duisburg. Upon graduation, and for the following 17 years, he performed static and dynamic analyses on steel structures for bridges and power plants, eventually specializing in damage analysis and refurbishment. Since 1990, he has been vastly involved in the area of plain bearings, focusing on metallurgy, new technologies, and damage analysis. He first worked at Th. Goldschmidt AG, Essen and, from 1998 on, at ECKA Granulate GmbH, where he became the head of the company's Plain Bearing Technology unit.

In 2002, he founded the VDMA (Verband Deutscher Maschinen und Anlagenbaun) working group on plain bearings, having also become its chairman.

His international experience, both at the manufacturing and end-user levels, provides him with deep insight into the challenges faced by those in the same field.

Mr. Koring's expertise presents us with a refreshing and unique confluence of theoretical and practical knowledge so hard to find in an era of specialists.